How to Effectively Assess the Impact of Non-Lethal Weapons as Intermediate Force Capabilities

KRISTA ROMITA GROCHOLSKI, SCOTT SAVITZ, JONATHAN P. WONG, SYDNEY LITTERER, RAZA KHAN, MONIKA COOPER

Prepared for the Joint Intermediate Force Capabilities Office
Approved for public release; distribution unlimited

NATIONAL DEFENSE RESEARCH INSTITUTE

For more information on this publication, visit **www.rand.org/t/RRA654-1**.

About RAND

The RAND Corporation is a research organization that develops solutions to public policy challenges to help make communities throughout the world safer and more secure, healthier and more prosperous. RAND is nonprofit, nonpartisan, and committed to the public interest. To learn more about RAND, visit www.rand.org.

Research Integrity

Our mission to help improve policy and decisionmaking through research and analysis is enabled through our core values of quality and objectivity and our unwavering commitment to the highest level of integrity and ethical behavior. To help ensure our research and analysis are rigorous, objective, and nonpartisan, we subject our research publications to a robust and exacting quality-assurance process; avoid both the appearance and reality of financial and other conflicts of interest through staff training, project screening, and a policy of mandatory disclosure; and pursue transparency in our research engagements through our commitment to the open publication of our research findings and recommendations, disclosure of the source of funding of published research, and policies to ensure intellectual independence. For more information, visit www.rand.org/about/principles.

RAND's publications do not necessarily reflect the opinions of its research clients and sponsors.

Published by the RAND Corporation, Santa Monica, Calif.
© 2022 RAND Corporation
RAND® is a registered trademark.

Library of Congress Cataloging-in-Publication Data is available for this publication.
ISBN: 978-1-9774-0857-0

Cover: Martin Wright/U.S. Navy.

About This Report

A key issue regarding non-lethal weapons (NLWs) is how to assess their tactical, operational, and strategic impact.[1] A nuanced understanding of NLWs' impact in various contexts is needed to inform their development and mainstream integration into overall DoD capabilities. This report describes how to effectively evaluate the impact of NLWs. This includes the development of a logic model structure that links NLW activities to outputs, higher-level outcomes, and ultimate strategic goals, including identification and evaluation of various metrics associated with the elements of that logic model. This report also includes results from analysis of the logic model and recommendations on NLW applications in support of DoD strategic goals. This work was undergirded by the development and analysis of diverse vignettes for NLW usage, together with numerous interviews and analyses of documents.

This research was sponsored by the Joint Intermediate Force Capabilities Office and conducted within the Navy and Marine Forces Center of the RAND National Security Research Division (NSRD), which operates the National Defense Research Institute (NDRI), a federally funded research and development center sponsored by the Office of the Secretary of Defense, the Joint Staff, the Unified Combatant Commands, the Navy, the Marine Corps, the defense agencies, and the defense intelligence enterprise. The research reported here was completed in August 2021 and underwent security review with the sponsor and the Defense Office of Prepublication and Security Review before public release.

For more information on the RAND Navy and Marine Forces Center, see www.rand.org/nsrd/nmf or contact the director (contact information is provided on the webpage).

---

[1] NLWs are a subset of Intermediate Force Capabilities (IFCs), a term introduced in 2020 by the U.S. Department of Defense (DoD) Non-Lethal Weapons Program, which is currently the Joint Intermediate Force Capabilities Office (JIFCO). IFCs encompass NLWs and a range of additional capabilities and technologies that cause less-than-lethal effects (see Susan LeVine, "Beyond Bean Bags and Rubber Bullets: Intermediate Force Capabilities Across the Competition Continuum," *Joint Forces Quarterly*, No. 100, First Quarter 2021, pp. 19–24).

Acknowledgments

We would like to thank the many people who shared their time and insights to help us conduct this research, representing the following organizations:

- Army Nonlethal Scalable Effects Center
- Irregular Warfare and Security Force Assistance, Force Modernization Proponent, Mission Command Center of Excellence, Army Combined Arms Center
- U.S. Army Security Assistance Training Management Organization
- U.S. Army Special Operations Command, G-9
- U.S. Coast Guard Pacific Command PAC-3
- U.S. Coast Guard CG-7211
- U.S. Coast Guard CG-MLE-2
- U.S. Air Force Security Forces Center
- U.S. Air Forces Europe A4/A4S
- U.S. Air Forces Pacific PACAF/A4S
- U.S. Air Forces Central Command Force Protection
- 354th Security Forces Squadron
- 3d Marine Division
- International and Operational Law Branch, Headquarters U.S. Marine Corps, Judge Advocate Division
- Headquarters, U.S. Marine Corps, Strategic Initiatives Group
- Marine Corps Warfighting Lab
- Naval Sea Systems Command 06, Expeditionary Missions Program Office (PMS 408)
- Navy Expeditionary Combat Command
- Office of the Secretary of Defense for Policy, Special Operations and Low-Intensity Conflict, Stability and Humanitarian Affairs
- Office of the Undersecretary of Defense for Research and Engineering
- Office of the Undersecretary of Defense for Acquisition and Sustainment and Office of the Assistant of Secretary of Defense for Acquisition, Platform and Weapon Portfolio Management
- NATO Counter-Improvised Explosive Device Center of Excellence
- ALEX–Alternative Experts Inc.
- Booz Allen Hamilton.

We would like to thank the Joint Intermediate Force Capabilities Office (JIFCO) for sponsoring this work. We received copious guidance and support from our action officers, Susan LeVine, S. Patrick Sweeney, and Col Wendell B. Leimbach, Jr. (the head of JIFCO). We would also like to thank the unit managers who oversaw this research, Paul DeLuca, Brendan Toland, Joel Predd, and Yun Kang, who helped to guide this work and provide valuable insights. Our colleagues, Ben Connable and James Kallimani, provided additional informal insights and knowledge. Finally, we gratefully acknowledge the feedback provided by John Aho and Michelle Ziegler during the review process, which helped to greatly improve the report and make it more accessible to wider audiences.

Summary

Background

The U.S. Department of Defense (DoD) is employing or developing various non-lethal weapons (NLWs), including acoustic hailers, laser dazzlers, flash-bang grenades, blunt-impact munitions (e.g., rubber bullets), tasers, pepper balls, an Active Denial System (ADS) that emits millimeter-wave energy to cause a temporary heating sensation, microwave-emitting technologies that disable vehicles and vessels, and vessel-stopping technologies that entangle or foul propellers.[1] By constraining other parties' courses of action without inflicting lethal force, NLWs can help achieve military ends while avoiding collateral damage. The importance of these capabilities may expand with increasing competition short of war (the *gray zone*), because they can help demonstrate resolve while mitigating the risks of unwanted escalation.[2]

---

[1] NLWs are a subset of Intermediate Force Capabilities (IFCs), a term introduced in 2020 by DoD Non-Lethal Weapons Program, which has been renamed the Joint Intermediate Force Capabilities Office (JIFCO). IFCs encompass NLWs and a range of additional capabilities and technologies that cause less-than-lethal effects (see Susan LeVine, "Beyond Bean Bags and Rubber Bullets: Intermediate Force Capabilities Across the Competition Continuum," *Joint Forces Quarterly*, No. 100, First Quarter 2021, pp. 19–24).

[2] According to a previous RAND report on the topic, "The gray zone is an operational space between peace and war, involving coercive actions to change the status quo below a threshold that, in most cases, would prompt a conventional military response, often by blurring the line between military and nonmilitary actions and the attribution for events" (see Lyle J. Morris, Michael J. Mazarr, Jeffrey W. Hornung, Stephanie Pezard, Anika Binnendijk, and Marta Kepe, *Gaining Competitive Advantage in the Gray Zone, Response Operations for Coercive Aggression Below the Threshold of Major War*, Santa Monica, Calif.: RAND Corporation, RR-2942-OSD, 2019, p. 8.

Purpose

A key challenge with respect to NLWs is how to evaluate their tactical, operational, and strategic impact to better inform decisionmaking throughout DoD regarding their development, acquisition, integration into military forces, and usage in diverse contexts. Whereas many DoD systems are assessed in terms of their ability to contribute to lethality or destruction, systems that deliberately aim to limit their effects require a different approach to evaluating their impact. Given this, the Joint Intermediate Force Capabilities Office (JIFCO) asked the RAND Corporation to develop a methodology to evaluate the tactical, operational, and strategic impact of these systems, particularly the strategic impact, given that is less intuitive.[3] The purpose of this report is to describe the results of that study on how to effectively measure the impact of NLWs.

Methodology

We conducted an extensive literature review on NLWs, reviewing more than 150 documents while also conducting 36 interviews with NLW stakeholders within and beyond DoD. Using the analysis of this information, we iteratively developed a structure called a *logic model* that links NLWs to DoD strategic goals. This logic model was refined based on further feedback from stakeholders to ensure accuracy, then used as the basis for identifying a series of metrics to measure each element within the logic model. We then developed a series of diverse vignettes to explore the degree to which the metrics were likely to be useful in characterizing the elements of the logic model in specific contexts. We also used thematic analysis to draw additional insights from the interviews, then developed overarching recommendations regarding how to assess and communicate the impact of NLWs.

---

[3] Obviously, we are not characterizing the immediate tactical effects of the weapons, which is part of the research, development, testing, and evaluation process.

Developing a Logic Model to Characterize NLWs

A logic model is a structured way to characterize how systems, processes, organizations, or other entities support goal achievement. Although there are many styles of logic models, our version describes how the *inputs* that enable NLW usage are used to conduct *activities* that contribute directly to *outputs*, then to higher-level *outcomes*, and, ultimately, to departmental-level *strategic goals*. Figure S.1 shows the logic model that we developed for this study. We then characterized the strength of the connections between elements in adjacent levels of the logic model. These connections helped illuminate which logic model elements are most relevant to DoD strategic goals.

Identifying Metrics to Evaluate the Logic Model

We identified at least one metric for each of the 29 elements of the logic model in the activity, output, and outcome categories, for a total of 97 unique metrics overall. Some of these metrics applied to more than one element of the logic model, resulting in 115 instantiations of metrics being applied to individual elements. We did not identify input metrics, which were either well-established technical specifications or were not relevant to measuring the impact of NLWs. Likewise, identifying metrics for DoD-wide strategic goals was well beyond the study's scope and irrelevant to evaluating NLWs' impact.

Developing Vignettes

To relate the logic model and its metrics to real-life events and to evaluate the utility of those metrics in specific contexts, we developed and analyzed a set of 13 vignettes in which NLWs might be used. These vignettes also helped us to corroborate the previously cited hypothesis, that NLWs are potentially useful in a wider range of tactical situations than those that they are primarily used in today. We developed vignettes based partly on past real-world events that included variability in terms of whether the adversary sought to escalate the situation, whether U.S. forces could feasibly withdraw,

FIGURE S.1
Logic Model for NLWs

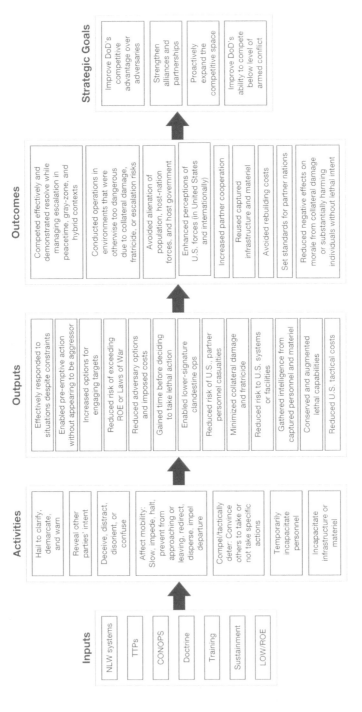

NOTES: CONOPS = concept of operations; LOW=Laws of War; ROE = rules of engagement; TTPs = tactics, techniques, and procedures.

and whether narrative surrounding the incident was stable. These vignettes also encompassed all of the services, the air, land, and maritime domains, and all geographic combatant commands. Examples included countering a small boat approaching a U.S. destroyer near the Strait of Gibraltar, addressing a motorized confrontation with Russian military contractors in Syria, dispersing demonstrators blocking entry to an air base in Palau during a conflict, managing a maritime standoff with Russian vessels in the Arctic, countering aggressive behavior in the air by Chinese aircraft, conducting a hostage rescue by special forces in Somalia, and countering self-styled domestic militias threatening humanitarian assistance and disaster response in Louisiana and Mississippi.

Evaluating Metrics in the Context of Vignettes

Once we developed our set of vignettes, we explored them using the logic model and associated metrics. We identified NLWs that would be applicable in the vignette, then assessed which elements of the logic model were relevant to the vignette. For all of the relevant elements, we collectively evaluated each of their metrics in the context of the vignette, ascertaining how well they measured the element and how consistently, easily, and quickly measurements were made.[4] Each metric was assigned a value along a three-point scale (low, medium, or high) for each of these criteria. Our subsequent analysis of these ratings revealed that most of the metrics scored well across all of these categories—although, on average, only about half of the metrics were applicable in any given vignette.

---

[4] Scott Savitz, Miriam Matthews, and Sarah Weilant, *Assessing Impact to Inform Decisions: A Toolkit on Measures for Policymakers*, Santa Monica, Calif.: RAND Corporation, TL-263-OSD, 2017.

Conclusions

Results from the Logic Model and Evaluated Metrics

Using our assessment of the connections between elements in the logic model, we were able to generate a few conclusions. First, the strong connections between different levels of the logic model illustrate which logic model elements are most relevant to DoD strategic goals. All seven of the activities in Figure S.1 ultimately have strong connections to the strategic goals, as do nine of the 13 outputs, and five of the nine outcomes. These elements (listed in Table S.1) and their associated metrics can be used to make the strongest case for the strategic impact of NLWs at a DoD-wide level.

TABLE S.1

Elements of the Logic Model with Strong Connections to DoD Strategic Goals

Activities	Outputs	Outcomes
• Hail to clarify, demarcate, and warn • Reveal other parties' intent • Deceive, distract, disorient, or confuse • Affect mobility (i.e., slow, impede, halt, prevent from approaching or leaving, redirect, disperse, impel departure) • Compel or tactically deter (i.e., persuade others to take or not take specific actions) • Temporarily incapacitate personnel • Incapacitate infrastructure or materiel	• Effectively responded to situations despite constraints • Enabled pre-emptive action without appearing to be aggressor • Increased options for engaging targets • Reduced risk of exceeding ROE or Laws of War • Reduced adversary options and imposed costs • Gained time before deciding to take lethal action • Enabled lower-signature clandestine operations • Reduced risk of U.S. and partner personnel casualties • Minimized collateral damage and fratricide	• Competed effectively and demonstrated resolve while managing escalation in peacetime, gray-zone, and hybrid contexts • Conducted operations in environments that were otherwise too dangerous due to collateral damage, fratricide, or escalation risks • Avoided alienation of population, host-nation forces, and host government • Enhanced perceptions of U.S. forces (in the United States and internationally) • Increased partner cooperation

Second, when we examined the metrics developed for the elements in each level of the logic model, a few trends emerged: (1) activity metrics primarily relate to which people or systems are affected by NLW usage and how well they respond to NLWs; (2) output metrics generally relate to providing the user with more time and options, curtailing the adversary's options, and reducing tactical risks; and (3) outcome metrics most often relate to reducing strategic and operational risks, influencing perceptions, maintaining morale, and reducing costs.

Finally, when applying the logic model and metrics to the vignettes, our analysis also revealed which NLWs were generally the most applicable to the range of contexts encompassed by our vignettes. We found that the IFCs that were particularly versatile were acoustic systems and laser dazzlers used to hail, deceive, distract, disorient, or confuse; and an ADS used to provide focused, discriminating effects that can tactically deter, deny access, or cause individuals to depart.

Combining this information could help JIFCO and other stakeholders structure discussions of how NLWs affect DoD's ability to achieve its tactical, operational, and strategic aims. For example, the direct tactical impact of NLW usage in a gray-zone encounter could affect another party's mobility: A ship's pilot, subjected to intense glare from a laser dazzler, chooses to divert its course away from the confrontation. The operational impact is that the United States has demonstrated resolve while managing escalation. Meanwhile, the strategic impacts include helping to compete below the level of armed conflict and proactively expanding the competitive space.

Themes Identified in Interviews

Much of what we learned about NLWs was gained from conducting 36 interviews of experts and stakeholders across 25 organizations spanning three broad categories: technologists involved in NLW development, policy-related personnel who provide resources and govern NLW usage, and operators who ultimately employ NLWs. Four key themes emerged from our analysis:

1. **Cultural and resource issues are the greatest challenges to NLW adoption.** Cultural issues primarily related to a reticence to embrace

NLWs even when doctrine and policy allowed for their use. This reticence often related to potential users having little confidence in NLWs working as intended, not seeing them as useful compared with lethal capabilities, or not fully understanding the effects of NLWs. In terms of resource challenges, interviewees highlighted that a lack of NLW availability and competing training demands often forced them to de-emphasize NLWs even when they might have been useful.

2. **NLWs are often perceived as burdensome** to the point that they are not carried into operational engagements due to logistical concerns and constraints.

3. **Challenges interact and reinforce each other.** An example of this is that commands with little familiarity with NLWs tend to discount their utility, so they limit the extent of NLW training and usage, which reinforces that unfamiliarity.

4. **Opportunities for additional NLW usage are not widely recognized.** For example, interviewees generally had little to say about the potential applicability of NLWs in strategic or great-power competition, beyond limited perception of NLW usage in gray-zone situations.

Recommendations

Leveraging the results of our analysis using the logic model, metrics, and series of vignettes, we recommend that JIFCO take a couple of natural next steps.

1. **Present and discuss the logic model in various forums, including with senior leaders, to convey how NLWs contribute to DoD strategic goals.** The logic model provides concrete descriptions of activities and relationships that have often been superficially or incompletely understood. As DoD continues to shift its focus toward competition with China and Russia, the logic model and exploratory vignettes make it clear how NLWs can contribute to that competition, including by explicitly linking NLWs to strategic goals from

the unclassified summary of the National Defense Strategy, helping to counter some of the misperceptions and misunderstandings about NLWs that interviews revealed.

2. **Work with the services to collect data that can be used to evaluate the impact of NLWs by providing values for the metrics.** The values of these metrics can also be measured in real-world operations and potentially also in live exercises or wargames. Metrics that relate to outputs and outcomes that have strong links to strategic goals, are relevant to a range of vignettes, and are easy to measure should be prioritized.

The study also revealed a host of issues that inhibit the use of NLWs. Many of these relate to perceptions of NLWs as burdensome, a lack of awareness regarding their prospective utility, a lack of adequate unit-level training and integration into TTPs, and misunderstood or ambiguous policies. There are also widespread negative perceptions of NLWs, including views that NLWs are more harmful than lethal weapons.

To overcome these factors, there are four main approaches that JIFCO should undertake.

1. Work with the services and other DoD stakeholders to ensure that policies and concepts of operations are consistent and clearly understood.

2. Collaborate with the Joint Chiefs of Staff (JCS) J7 on joint training standardization regarding NLWs to ensure that services provide thorough unit training with NLWs and that NLWs are tightly integrated into units' TTPs. Although the services direct their own training, and JNLWO lacks authority in this area, the JCS J7 can help shape NLW training standards across DoD. JIFCO can also work directly with the services or particular units, given its interest in doing so, to ensure that units are adequately trained for NLW usage.

3. Shape perceptions within the military via explanations using the logic model, exploration of vignettes, demonstrations both in live exercises and in wargames, and the use of data sets to measure NLWs' impact (once those become available).

4. Future NLWs should be designed from the outset to minimize the aspects of them that contribute most to perceived and actual burdens. JIFCO should support the development of NLWs (as part of the acquisition process) that are designed to be easy to use, low maintenance, and have reduced space, weight, and power requirements, even at the expense of other desired attributes, to make them more attractive to future users.

The study's finding that acoustic systems, laser dazzlers, and an ADS are especially versatile in a variety of scenarios could contribute to these systems being used in an array of contexts that might not previously have been fully realized.

Closing Remarks

In this report, we describe how the tactical, operational, and strategic impact of NLWs can be characterized using a logic model and a set of associated metrics. This factor clarifies how NLWs relate to DoD strategic goals and—in tandem with observations from interviews about how NLWs are perceived—how it facilitates better communication within DoD regarding how these systems can be better integrated into operations. The identification and characterization of the metrics also lay the groundwork for data collection that can be used to further evaluate the impact of NLWs at multiple levels, which in turn can shape their usage in ways that enhance their contributions to DoD effectiveness.

Contents

Figures

Tables

CHAPTER ONE

Introduction

Background

The U.S. Department of Defense (DoD) is employing or developing various non-lethal weapons (NLWs), including acoustic hailers, laser dazzlers, flash-bang grenades, blunt-impact munitions (e.g., rubber bullets), tasers, pepper balls, the Active Denial System (ADS) that emits millimeter-wave energy to cause a temporary heating sensation, microwave-emitting technologies that disable vehicles and vessels, and vessel-stopping technologies that entangle or foul propellers. NLWs are a subset of Intermediate Force Capabilities (IFCs), a term introduced in 2020 by the DoD Joint Non-Lethal Weapons Program, now termed the Joint Intermediate Force Capabilities Office (JIFCO). IFCs encompass NLWs and a variety of additional capabilities and technologies that cause less-than-lethal effects.[1]

By constraining other parties' courses of action without inflicting lethal force, NLWs can help achieve military ends despite apprehensions about collateral damage or escalation. Less-than-lethal capabilities have previously primarily been used in a law enforcement context; however, their importance may expand with increasing competition short of war (the gray zone) because, in a standoff, they can help communicate and demonstrate

---

[1] See Susan Levine, "Beyond Bean Bags and Rubber Bullets: Intermediate Force Capabilities Across the Competition Continuum," *Joint Forces Quarterly*, No. 100, First Quarter 2021, pp. 19–24.

resolve while mitigating the risks of unwanted escalation.[2] For example, if Chinese or Russian forces confrontationally approach their U.S. counterparts with the intent of compelling them to depart from an area, NLWs can offer a means of countering these aggressive behaviors without the use of lethal force.

A key challenge with respect to NLWs is how to evaluate their tactical, operational, and strategic impact. Many DoD systems are assessed in terms of their ability to contribute to lethality or destruction: for example, the extent of damage that a munition can inflict against a fixed target, or the size of the area within which it can inflict a given degree of damage. Systems that deliberately aim to limit their own effects require a different approach to evaluating their impact.

Given this, JIFCO asked the RAND Corporation how best to evaluate the tactical, operational, and strategic impacts of these systems to better inform decisionmaking throughout DoD regarding their development, acquisition, integration into military forces, and usage. JIFCO, previously known as the Joint Non-Lethal Weapons Directorate, serves as the management office for the DoD Non-Lethal Weapons Program, for which the Commandant of the Marine Corps is the Executive Agent. It is important for JIFCO, stakeholders throughout the services, and DoD to be able to define what value NLWs can provide to ascertain how to shape their development and employment.

Defining NLWs

As stated earlier, the term *IFC* was introduced in 2020 to better reflect the range of capabilities and technologies available for enacting less-than-lethal effects in contemporary operating environments well beyond their legacy association with law enforcement. At the time of this writing, this term

---

[2]   According to a previous RAND report on the topic,

> The gray zone is an operational space between peace and war, involving coercive actions to change the status quo below a threshold that, in most cases, would prompt a conventional military response, often by blurring the line between military and nonmilitary actions and the attribution for events (Lyle J. Morris, Michael J. Mazarr, Jeffrey W. Hornung, Stephanie Pezard, Anika Binnendijk, and Marta Kepe, *Gaining Competitive Advantage in the Gray Zone, Response Operations for Coercive Aggression Below the Threshold of Major War*, Santa Monica, Calif.: RAND Corporation, RR-2942-OSD, 2019, p. 8).

remains predecisional, and a standardized definition has not yet been codi-fied in doctrine. Our focus is on NLWs, which are a subset of IFCs. In the absence of clear definitions for either term at present, we developed a non-doctrinal definition of NLWs for this report, drawing on the definition of NLWs from DoD Directive 3000.03E:[3]

> Systems and capabilities that can be used in all phases of conflict to stop, deter, deny, delay or temporarily incapacitate targeted person-nel and materiel by producing predictable, immediate effects that are intended to be reversible and minimize unnecessary destruction and loss of life.

The phrase *all phases*, taken from the DoD directive, is a reminder that these systems can be used in Phase 0 of military conflict (shaping of the environment) and in the inter-war phase, in which the usage of lethal sys-tems is severely restricted.[4]

Systems falling under this definition vary greatly in terms of how they achieve the desired effects. Some examples of such systems include the following:

- **Acoustic systems.** The acoustic hailing device (AHD) can be used to communicate orally at long distances—e.g., to tell someone to back away.[5] The experimental concept Laser-Induced Plasma Effects may use lasers to create a distant ball of plasma that can create sounds, including human speech, to persuade people to alter their movements or behavior.[6]

---

[3] DoD Directive 3000.03E, *DoD Executive Agent for Non-Lethal Weapons (NLW), and NLW Policy*, Washington, D.C., April 25, 2013, Incorporating Change 2, August 31, 2018.

[4] Joint Publication 5-0, *Joint Planning*, Washington, D.C.: Joint Chiefs of Staff, Decem-ber 1, 2020.

[5] JIFCO, DoD, Non-Lethal Weapons Program, "Acoustic Hailing Devices Fact Sheet," November 16, 2018a.

[6] Patrick Tucker, "The US Military Is Making Lasers Create Voices out of Thin Air," *Defense One*, March 2, 2018.

- **Laser dazzlers.** These include the currently fielded Ocular Interrupter (OI) and developmental Long-Range Ocular Interrupter (LROI), both of which create intense glare that prevents people from being able to perceive their environment well but without any permanent effects (in keeping with the Protocol on Blinding Laser Weapons).[7] They can also be used to gain someone's attention (hail) at long ranges.
- **Integrated-effects systems.** The still-in-development Escalation of Force (EoF) Common Remotely Operated Weapons Station (CROWS) includes acoustic, light, and laser dazzling capabilities.[8]
- **Flash-bang grenades.** These create a burst of intense light and sound to distract and temporarily incapacitate individuals.[9]
- **Blunt impact munitions.** These include rubber bullets, beanbag rounds, grenades that disperse rubber pellets, and other systems intended to strike individuals to temporarily incapacitate them while limiting the scope of permanent injuries.[10]
- **Electro-muscular incapacitation systems.** These short-range devices use an electrical current to induce incapacitating muscle contractions. Tasers allow a modest degree of standoff distance.[11]
- **Riot control agents.** These are non-lethal chemical irritants, such as pepper spray and tear gas, that are typically reserved for law enforcement and crowd-control situations. The Chemical Weapons Convention precludes their use in warfare; however, U.S. interpretation and ratification of the Chemical Weapons Convention allows for very lim-

---

[7] JIFCO, DoD, Non-Lethal Weapons Program, "Non-Lethal Optical Distracters Fact Sheet," May 2016a; and U.S. Department of Defense, Office of General Counsel, *Department of Defense Law of War Manual*, Washington, D.C., June 2015, updated December 2016, pp. 411–412.

[8] JIFCO, DoD, Non-Lethal Weapons Program, *DoD Non-Lethal Capabilities: Enhancing Readiness for Crisis Response Annual Review*, Quantico, Va.: Non-Lethal Weapons Program, 2015; and U.S. Department of Defense, Non-Lethal Weapons Program, *Intermediate Force Capabilities: Bridging the Gap Between Presence and Lethality, Executive Agent's Planning Guidance 2020*, March 2020a.

[9] JIFCO, DoD, Non-Lethal Weapons Program, 2015.

[10] JIFCO, DoD, Non-Lethal Weapons Program, 2015.

[11] JIFCO, DoD, Non-Lethal Weapons Program, "Human Electro-Muscular Incapacitation FAQs," webpage, undated-b.

ited use as delineated in Presidential Executive Order 11850.[12] Pepper spray can be used at short ranges, while pepper balls can be used to disperse effects over wider areas.[13]

- **Millimeter-wave systems.** The Active Denial System (ADS) emits a focused beam of millimeter-wave energy to safely and rapidly cause a temporary, immediately reversible heating sensation to deny personnel access to an area or encourage them to move. A developmental version, ADS Solid State, will reduce system weight and power requirements to improve mobility.[14]

- **Microwave systems.** JIFCO is also completing prototype development and assessment for systems that temporarily interfere with vehicle electronics using high-power microwaves, including short- and long-range Radio Frequency Vehicle Stoppers (RFVSs) for stopping land-based vehicles and the Vessel Incapacitating Power Effect Radiation (VIPER) system for maritime use.[15] Similar systems are envisioned to counter unmanned aerial vehicles (UAVs).[16]

---

[12] Gerald Ford, "Renunciation of Certain Uses in War of Chemical Herbicides and Riot Control Agents," Washington, D.C.: Executive Office of the President, Executive Order 11850, April 8, 1975.

[13] JIFCO, DoD, Non-Lethal Weapons Program, "Oleoresin Capsicum Dispensers," webpage, undated-c; and JIFCO, DoD, Non-Lethal Weapons Program, "Variable Kinetic System (VKS) Non-Lethal Launcher and U.S. Coast Guard Pepperball Launcher Systems," webpage, undated-d.

[14] JIFCO, DoD, Non-Lethal Weapons Program, "Active Denial System FAQs," webpage, undated-a; and Susan LeVine, *The Active Denial System: A Revolutionary, Nonlethal Weapon for Today's Battlefield*, Washington, D.C.: Center for Technology and National Security Policy, National Defense University, 2009.

[15] Jamal Beck, "New Vehicle Stopper Trials Underway at Tinker Air Force Base," press release, Joint Intermediate Force Capabilities Office, U.S. Department of Defense, Non-Lethal Weapons Program, August 15, 2018; JIFCO, DoD, Non-Lethal Weapons Program, "Vessel-Stopping Prototype," November 16, 2018d; and JIFCO, DoD, Non-Lethal Weapons Program, "Radio Frequency Vehicle Stopper," November 16, 2018b.

[16] Other nations have sometimes used intense bursts of microwaves to target people—e.g., they have been used to attack U.S. personnel in Cuba and Russia, causing brain damage (see Julia Ioffe, "The Mystery of the Immaculate Concussion," *GQ*, October 19, 2020).

- **Mechanical vehicle/vessel-stopping technologies.** The Single Net Solution–Remote Deployment Device (SNS-RDD) consists of a spiked net deployed to stop land-based vehicles, and the Pre-Emplaced Vehicle Stopper (PEVS) injects electricity into a vehicle to damage its electronics. The Maritime Vessel Stopping Occlusion Technologies (MVSOT) include drogue lines (which tangle the propellers) and occlusion technologies (which coat propellers to reduce efficiency and effectiveness).[17]

Some non-lethal applications of cyber and electronic warfare may appear to fall within the definition given above but are not overseen by JIFCO in accordance with DoD Directive 3000.03E.[18] As a result, they were excluded from the scope of this study.

Generally, much of the services' usage of NLWs has been for military police and security forces, garrison law enforcement, crowd control, or defense of fixed sites. (Some of these systems have also been used by civilian law enforcement agencies, both domestically and internationally.)[19] Using the literature review, we developed the hypothesis that NLWs could be useful in a range of other situations, such as standoffs in the context of the gray zone of competition with forces from rival nations. We explored and corroborated this hypothesis throughout the study, demonstrating it in the context of 13 vignettes.

---

[17] Army Nonlethal Scalable Effects Center, "Army Nonlethal Weapons APBI," presentation, September 21, 2017; Katherine Mapp, "Promising New Tool Protects Ships, Sailors," Naval Surface Warfare Center Panama City Division, Public Affairs, November 21, 2019; Nathan Gain, "US Navy Lab Investigates Innovative Non-Lethal Boat Stopping Technology," *Naval News*, November 25, 2019; and JIFCO, DoD, Non-Lethal Weapons Program, "Single Net Solution with Remote Deployment Device," November 16, 2018c.

[18] Systems emitting millimeter waves and microwaves use the electromagnetic spectrum, but these differ qualitatively from the more-sophisticated jamming, spoofing, and manipulation involved in most electronic warfare (EW), so they are included in this study.

[19] In the context of civilian law enforcement, these systems can be controversial, and debate over their usage in such contexts is (at the time of this writing) intensified by wider disagreements over policing. However, the issue of whether, how, and when such systems should be used in civilian law enforcement is beyond the scope of this study, which focuses exclusively on military use of NLWs.

Purpose of This Study

The purpose of this study is to characterize how best to measure the potential tactical, operational, and strategic impact of NLWs, ascertaining how they contribute to overarching goals. Clear assessments of the impact of NLWs at multiple levels can provide insights that can inform their development and acquisition, how they are integrated into military operations, what NLW CONOPS to use in different contexts, and what TTPs to employ with NLWs, particularly in confrontations with near-peer adversaries or when collateral damage could have grave strategic effects.

Methodology and Report Structure

Here, we present a brief outline of the methodology, with more details provided in the subsequent chapters that describe its various parts and the associated analyses. We began by qualitatively characterizing how the direct effects of NLWs contribute, via a series of steps, to high-level DoD goals. To do this, we conducted a literature review and a series of interviews. We reviewed more than 150 documents, including DoD policy and doctrine, budget documents, RAND reports, academic journals, and news articles. We also conducted 36 interviews with stakeholders from 25 different organizations, across and beyond DoD. We used the information gleaned from these sources and a series of internal workshops to iteratively develop a logic model that links the activities NLWs perform, the direct outputs of those activities, higher-level outcomes, and DoD strategic goals. This logic model was refined based on further feedback from stakeholders to ensure accuracy. The methodology for developing the logic model is further described in Chapter Two. Additional information on the intensity of the relationships between elements of the logic model is included in Appendix A.

We then identified metrics that could be used to measure each element (item) within the logic model, based on analysis by individuals and the collective research team. We then developed a series of diverse vignettes to evaluate the degree to which the metrics were likely to be useful in characterizing the elements of the logic model in specific contexts; more information about how the vignettes were developed and the metrics evaluated appears in Chapter Three, while Appendix B provides descriptions of the metrics and the results of their evaluations, and Appendix C includes detailed information about the vignettes. In parallel, we used thematic analysis of interview data to generate additional insights, as will be further described in Chapter Four. The interview protocols used in these interviews are provided in Appendix D. Finally, we developed a set of overarching assessments regarding how to evaluate and communicate the impact of NLWs, captured in Chapter Five.

CHAPTER TWO

Developing a Logic Model to Characterize NLWs

In this chapter, we review the base of our framework for understanding the impacts of using NLWs. This takes the form of a logic model that links the inputs required to use NLWs to the higher-level impacts of their use. First, we describe the basic structure of a logic model and how we populated it. This is followed by a description of the logic model used to identify the metrics discussed in Chapter Three. We conclude by characterizing connections between individual elements at various levels of the logic model and offering some initial observations.

How a Logic Model Works

A logic model is a structured way to characterize how systems, processes, organizations, or other entities achieve their goals. Although there are many styles of logic models, the version that we are using describes how the *inputs* that enable NLW usage are used to conduct *activities* that contribute directly to *outputs*, then to higher-level *outcomes*, and ultimately, to departmental-level *strategic goals*, as shown in Figure 2.1.[1] The tactical impact of NLWs is captured at the activity level, while their operational impact is captured at the output level, and their strategic impact is captured at the outcome and strategic goal levels.

---

[1] Scott Savitz, Miriam Matthews, and Sarah Weilant, *Assessing Impact to Inform Decisions: A Toolkit on Measures for Policymakers*, Santa Monica, Calif.: RAND Corporation, TL-263-OSD, 2017.

FIGURE 2.1

NLW Logic Model Structure

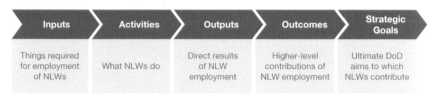

Inputs	Activities	Outputs	Outcomes	Strategic Goals
Things required for employment of NLWs	What NLWs do	Direct results of NLW employment	Higher-level contributions of NLW employment	Ultimate DoD aims to which NLWs contribute

SOURCE: Savitz, Matthews, and Weilant, 2017.

How the Logic Model Was Developed

The content and organization of our logic model was informed by a review of more than 150 documents related to NLWs and the various contexts in which they might be employed and by the 36 interviews we conducted with stakeholders across DoD, the U.S. Coast Guard, and other relevant parties. Our team conducted multiple workshops throughout the document review process to synthesize our findings into a list of NLW uses and impacts, which we organized into a logic model according to the basic framework shown in Figure 2.1. In each subsequent workshop, we iterated on the model using insights provided by additional literature.

After iterating on the logic model based on our literature review and interviews, we requested feedback on its content and organization from JIFCO and other subject-matter experts, further refining the model based on their responses. The framework presented here should be considered a living document, to be adapted as (1) new technologies and creative operational concepts expand the boundaries of what is possible and (2) improved data collection and analysis refine our understanding of these capabilities.

Contents of the Logic Model

In developing this logic model, we found that the potential impacts of NLWs are many and varied. This is unsurprising because the capabilities themselves are varied, relying on a diverse set of technologies and physical phenomena, such as sound, light, millimeter waves, blunt impact, and chemical irritation, as noted earlier in this chapter. Furthermore, they are meant for use in a wide range of situations against a variety of potential targets (i.e., individuals, crowds, vessels, vehicles).[2] Figure 2.2 shows the logic model used in the framework discussed in this report.

The elements, listed in the leftmost column of Figure 2.2, consist of tangible and intangible inputs required to employ NLWs effectively and appropriately in a given situation. Most obviously, NLW employment requires the physical systems and the means to sustain their use. In addition, DoD must determine how and when personnel can and should use NLWs, and personnel must understand both the operation of the physical systems and the broader guidelines for their use. Activities, listed in the second column from the left in Figure 2.2, consist of things NLWs do, such as hail, distract, incapacitate, or affect mobility (the last includes the concept of area denial, preventing access to an area). It is possible for NLWs to conduct multiple activities depending on how they are employed. For example, a laser dazzler could be used against the driver of a car approaching a checkpoint to warn the driver to slow down for the checkpoint and visually degrade or suppress if the person fails to do so. There is also some degree of overlap between the activities (e.g., tactically deterring someone from approaching also affects that person's mobility). The output elements, listed in the middle column of Figure 2.2, are the direct results of NLW employment. Outputs include affecting U.S. and adversary costs; reducing various forms of risk; increasing time, information, and options; and enabling effective action in various situations despite constraints. Outcomes, listed in the second column from the right in Figure 2.2, are one step further removed, consisting of high-level impacts to how and where the United States can operate, the broad costs incurred by U.S. operations, and the perceptions they create. Finally, strategic goals, listed in the rightmost column in Figure 2.2, are broad, department-wide goals set out by DoD leadership—specifically, the goals from the National Defense Strategy unclassified summary that NLWs can

---

[2] JIFCO, DoD, Non-Lethal Weapons Program, *Strategic Plan 2016–2025, Science & Technology, Joint Non-Lethal Weapons Program*, Quantico, Va., 2016b; and Richard L. Scott, "Nonlethal Weapons and the Common Operating Environment," *ARMY Magazine*, April 2010, pp. 21–26.

FIGURE 2.2
Logic Model for NLWs

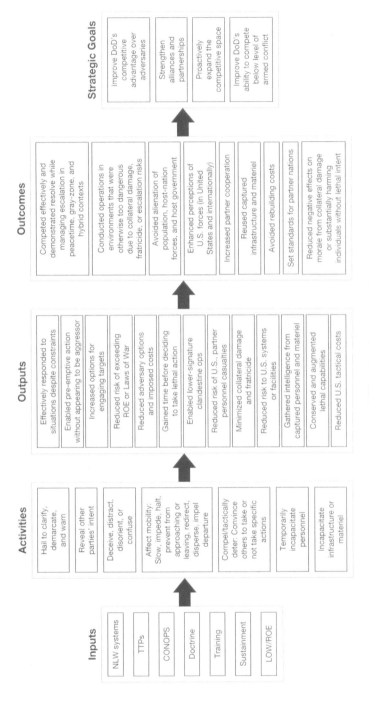

NOTES: CONOPS = concept of operations; LOW=Laws of War; ROE = rules of engagement.

help fulfill.[3] Although NLWs are obviously not wholly responsible for fulfillment of these goals, they can play a role in contributing to them. The next section discusses our work in making the relationships among different levels of the logic model more explicit.

Connections Among Elements of the Logic Model

After developing the logic model, which shows generally how sets of elements feed into one another, we developed a more-detailed mapping of the relationships between individual logic model elements. Because the logic model aims to describe the impacts of diverse capabilities used in a broad set of situations, we can form a more-coherent picture of the different mechanisms by which NLWs create higher-level impacts and determine which logic model elements are most relevant when specific goals or scenarios are considered.

In the absence of data quantifying the relationships among different elements of the logic model, we relied on the expertise of the project team to determine the strength of the connections among logic model elements, informed by our literature review and discussions with other subject-matter experts and later provided to stakeholders for feedback. We used a simple, three-point scale:

- 2: strong, unequivocal connection
- 1: limited, indirect, or conditional connection
- 0: no connection.

We characterized only connections between elements in adjacent levels of the logic model, assuming that relationships between elements in nonadjacent levels arose through a chain of direct relationships with elements at intermediate levels. We did not map inputs to activities in the figure because we assessed that all inputs contribute to all activities, and leaving out this all-to-all mapping enables the reader to focus on more-nuanced aspects of

---

[3] James Mattis, *Summary of the 2018 National Defense Strategy: Sharpening the American Military's Competitive Edge*, Washington, D.C.: U.S. Department of Defense, 2018.

the diagram. Figure 2.3 provides a visual representation of element connectivity in the logic model, with strong, unequivocal connections shown as dark, heavy lines, and more limited, indirect, or conditional connections shown as lighter, thinner lines. Although this diagram is somewhat dense, we highlight the key takeaways from it—which can be identified visually by looking holistically at patterns across the diagram rather than focusing on individual connections—below. A list of all mappings among the logic model elements can be found in Appendix A.

As Figure 2.3 shows, activity-to-output links are dense, reflecting the versatility of the activities in contributing to diverse outputs. On average, each activity is connected to 90 percent of the outputs and strongly connected to 64 percent of them. Output-to-outcome connections are less dense: Most outputs link to only a few outcomes (an average of 49 percent, with strong connections to only 23 percent, on average), and some outputs lack strong links to any outcomes. The outcome-to-strategic goal connections are still more sparse, with only five of the nine having strong links to at least two strategic goals and the other four having no strong links to any goals.

Using the connections depicted in Figure 2.3, we can see which logic model elements are most relevant to DoD's strategic goals. Figure 2.4 highlights the elements that are strongly connected to the strategic goals, either directly or via a chain of strong connections. All seven of the activities ultimately have strong connections to the strategic goals, as do nine of the 13 outputs and five of the nine outcomes. These elements make the strongest case for the strategic impact of NLWs at a DoD-wide level. This is not to say that other elements of the logic model do not have a substantial impact, only that their tactical and operational impacts do not relate as strongly to overall strategic goals.

FIGURE 2.3

Logic Model with Linkage Among Elements Shown

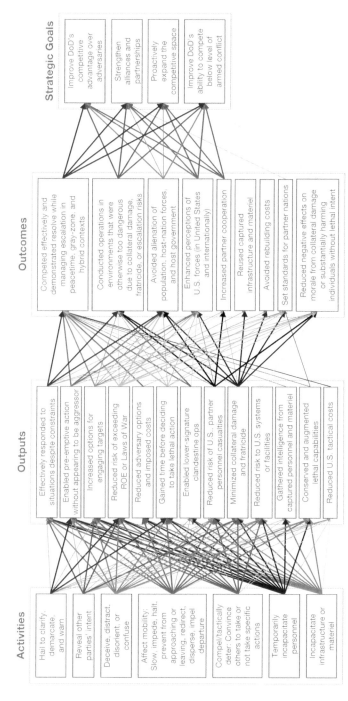

NOTE: Strong and limited connections are shown by thick and thin lines, respectively. Different line colors are used solely to improve clarity; all arrows emanating from a single element share the same color.

FIGURE 2.4

Logic Model Highlighting Elements Most Relevant to DoD's Strategic Goals

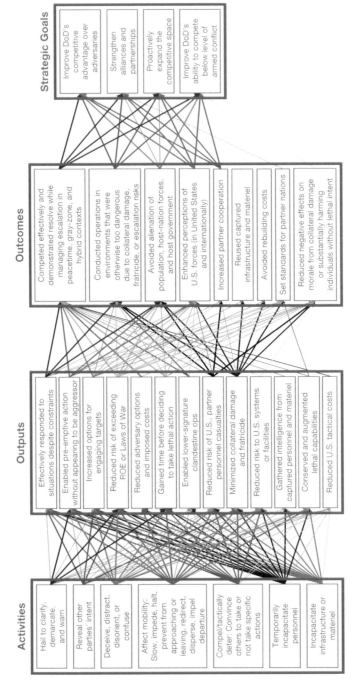

NOTE: Strategic goals and elements with strong connections to those goals (either directly or through other elements) are contained in dark blue boxes. Different line colors are used solely to improve clarity.

CHAPTER THREE

Identifying and Evaluating Metrics in the Context of Vignettes

Chapter Two provided an overview of the logic model, a framework for understanding the impacts of NLW usage. In this chapter, we present the metrics that we identified to measure the individual elements of that logic model; the values of these metrics could provide insights regarding the impact of NLWs. We then describe the methodology used for identifying these metrics and a brief overview of the characteristics of metrics by level of the logic model. Next, we discuss the method used to evaluate each of the identified metrics. We follow this with an explanation of the purpose of the vignettes, the process for developing them, and our related analyses. (More information about the vignettes is included in Appendix C.) Finally, we report the results of the analysis, including key observations regarding the metrics and how their values could be assessed through future data collection. Although we sought current data sets that might provide values for these metrics, none of the data sets that we were able to unearth—ranging from succinct tactical reports of NLW usage to indications of intrusions that garnered a response—had sufficient data to enable the values of these metrics to be calculated.

Identifying Metrics

The method used to identify the metrics was similar to the iterative process used for developing the logic model. Our team, in a series of internal workshops, went through each element in the activities, outputs, and outcomes levels of the logic model and discussed potential metrics that could be used to measure that element. Given the abstract and complex nature of some of the elements of the logic model, it was necessary to think creatively to develop a set of metrics that encompassed both quantitative and qualitative aspects. Once an initial set of metrics was developed, we considered them as a whole and made clarifying adjustments to create consistency across similar elements in the logic model and identified additional metrics to fill in any gaps. Ultimately, we identified 97 unique metrics to assess 29 logic model elements. Some of these metrics are applicable to multiple elements, resulting in 115 instantiations of metrics. The breakdown of the number of metrics by level of the logic model is included in Table 3.1, while Table 3.2 shows illustrative examples of metrics associated with the activities, outputs, and outcomes levels of the logic model. A complete list of metrics is included in Appendix B.

In general, good metrics were easiest to identify for activities, which primarily related to tangible items, such as which populations were affected by the use of NLWs and how they responded. It was harder to identify metrics for outputs, which generally related to providing the user with more time and options, curtailing the adversary's options, and reducing tactical risks. The hardest metrics to identify were those measuring outcomes, which related to more abstract phenomena (e.g., reducing strategic and operational risks, influencing perceptions, maintaining morale, and reducing costs).

Although we considered identifying input metrics, we found that our draft metrics for this level of the logic model were generally either well-established technical specifications (e.g., range and weight) or were not relevant to measuring NLW impacts, such as metrics associated with ROE. Therefore, we did not identify a set of metrics for the input level of the logic model. We also did not identify metrics associated with the strategic

TABLE 3.1
Numbers of Elements and Metrics

Element Type	Number of Elements	Number of Metrics
Activity	7	33
Output	13	42
Outcome	9	40
Total	29	115

TABLE 3.2

Examples of Metrics Associated with a Subset of Elements of the Logic Model

Element Type	Element Description	Metric
Activity	Temporarily incapacitate personnel	Percentage of targeted population incapacitated by NLW
		Percentage of encounters in which non-targeted population is incapacitated by NLW
		Timeline between NLW use and incapacitation
		Duration of incapacitation
Output	Effectively responded to situations despite constraints	Percentage of tactical encounters in which use of NLWs was permissible, but lethal force was not
		Whether NLWs are allowed by ROE (binary yes/no distinction)
		Degree to which targeted populations perceive NLWs as equivalent to lethal weapons
Outcome	Competed effectively and demonstrated resolve while managing escalation in peacetime, gray-zone, and hybrid contexts	Percentage of incidents using NLWs that resulted in unwanted escalation divided by percentage of incidents not using NLWs that resulted in unwanted escalation
		Percentage of particular peacetime/gray-zone/hybrid incidents in which NLWs were used
		Percentage of incidents in which NLWs were used and commanders perceived them as contributing effectively
		Degree to which targeted populations perceive NLWs as equivalent to lethal weapons

goals, because this is well beyond the scope of JIFCO's authority: Defining the best ways to measure the fulfillment of DoD-wide strategic goals from the National Defense Strategy would be a gargantuan task requiring the involvement of high-level stakeholders throughout DoD rather than a much smaller set of NLW stakeholders, and it also would not closely relate to the impacts of NLWs themselves.

Evaluating Metrics

We evaluated each metric (not the value of the metric, but the qualities of the metric itself) using four criteria that are widely used for this purpose, as described in an earlier RAND publication.[1] They are

- validity—how well the metric measures the element
- reliability—the degree to which multiple measurements will be consistent
- feasibility—how easily the measurement can be made
- timeliness—how quickly a measurement can be made.

The team evaluated each metric in the context of each of the vignettes that we explored, which we will describe next. For each metric, we began by ascertaining whether it was applicable in the context of that vignette (assigning it an N/A for *not applicable*, if not). If the metric was *applicable*, the team then collectively evaluated the metric relative to each of the four criteria, using a three-point scale (low, medium, high) for each. We will further explain the process and discuss the results of this analysis after we provide more information about the vignettes themselves.

Developing Vignettes

Rather than trying to analyze metrics in isolation, we ascertained that it was important to evaluate them in the context of vignettes that reflected a diverse range of use cases for NLWs. The use of vignettes also allowed us to test the hypothesis that NLWs are potentially useful in a wider range of tactical situations than they are primarily used in today. The settings of the vignettes are in the mid-2020s, allowing some NLW technologies that are

---

[1] Savitz, Matthews, and Weilant, 2017.

currently developmental to become fully operational, but the global situation is roughly the same as that at the time of this writing in 2021.

To implement this approach, we had to ensure that we could confidently enumerate all relevant types of vignettes. We turned to previous RAND research on scenario design, which highlighted two considerations. First, NLWs may be useful across the spectrum of conflict, so there is a need to distinguish between structural and proximate factors of conflict when designing scenarios.[2] Second, non-military factors need to be strengthened and characterized in greater fidelity to improve the quality of military and political strategic analysis.[3]

Using these factors, we developed three key design considerations that we varied to yield the relevant range of vignettes. The considerations, in detail, were as follows:

- Whether the adversary sought to escalate the situation. This consideration is important because it allows us to consider the effect of NLWs on a situation in which the eventual use of lethal force by the adversary was possible. This gives us some insight into whether NLWs contain inherent de-escalatory qualities, or whether the situation or context dominates.
- Whether U.S. forces could feasibly withdraw. U.S. withdrawal would inherently de-escalate a situation, and we sought to understand how NLWs could contribute to situations in which the United States would have to stay and work through a tactical situation to its conclusion, as compared with situations in which U.S. forces would have to withdraw.
- Whether the narrative surrounding the incident was stable. This consideration addressed the role of narrative, information, and disinformation in a particular vignette. This consideration allows us to bridge the relationships between structural and proximate factors of a conflict or tactical encounter with the non-military factors pertinent to our analyses. This is particularly important to test the hypothesis that

---

[2] Timothy R. Heath and Matthew Lane, *Science-Based Scenario Design: A Proposed Method to Support Political-Strategic Analysis*, Santa Monica, Calif.: RAND Corporation, RR-2833-OSD, 2019, pp. 11–16.

[3] Heath and Lane, 2019, p. 6.

NLWs are particularly useful in countering or addressing gray-zone tactics and operational approaches by adversaries. There, narratives play a key role, and understanding how a situation can be manipulated (or not) from a perspective of narrative is important.

Because these considerations are binary (e.g., a narrative is either stable or unstable), eight combinations are possible. We built vignettes around these combinations, basing them on contemporary and past events whenever possible. For example, consider the combination of an escalatory adversary, a situation in which U.S. forces can withdraw, and an unstable narrative. We saw that incidents between U.S. and Russian ground forces in Syria fit that combination, so we built a vignette around it.[4] Ultimately, we developed a set of 13 vignettes that encompassed all eight combinations, all of the services, all physical domains, and all geographic combatant commands. They also include instances of great-power competition with China and Russia, as well as possible terrorist threats, confrontations with hostile crowds (both in the United States and abroad), and other contexts. Obviously, these vignettes are illustrative rather than comprehensive: a variety of other use cases for NLWs, or variations on these, could readily be created. However, these capture some of the varied circumstances in which NLWs could be used in many different contexts. Our analysis of them confirmed a hypothesis that we had developed at the outset of the study, as mentioned in Chapter One—specifically that NLWs have potential applications beyond the well-known repertoire of crowd control and law enforcement. The 13 vignettes are briefly outlined in Table 3.3 and described in greater detail in Appendix C.

To give the reader a sense of what a vignette entails, we describe two interrelated ones here. In vignettes 7 and 8, the U.S. Air Force has set up an air base in Palau, about 500 miles east of the southern Philippines, from which it is conducting operations to counter aggression in the region. In vignette 7, an adversary's agents have paid members of the local population, including families with young children, to demonstrate in front of the base gates, preventing movement into or out of the base. The Air Force does not

---

4 Eric Schmitt, "U.S. Troops Injured in Syria After Collision with Russian Vehicles," *New York Times*, August 26, 2020.

want to hurt or arrest the demonstrators, and neither does the local police force, but they are unable to persuade them to disperse, and the obstruction of the gates is impeding the base's ability to operate effectively. For this reason, the Air Force is interested in employing NLWs to disperse the crowd.

In vignette 8, the adversary's agents have also distributed high-powered laser pointers to local adolescents and are paying them to aim those laser pointers at the cockpits of aircraft that are landing or taking off. The adolescents have been observed by the base's UAVs and from its air traffic control tower. The Air Force and local authorities do not want to harm the adolescents, but they do want to stop them from dazzling pilots, which can cause fatal crashes or long-term eye damage. As with the base demonstrators, NLWs can be used to disrupt the adolescents' activity without inflicting permanent harm.

In both vignettes, Air Force personnel are also acutely aware that injuring or killing unarmed members of the local community is both an ethical and a political concern, because the government of Palau could then come under intense popular pressure to deny the United States further access to the base. This is one of the goals of the agents behind both vignettes: provoking U.S. forces to take lethal action, or having Palauan forces do so on behalf of the United States, could result in loss of basing rights. Therefore, the ultimate opponent in both vignettes is seeking escalation, as noted in Table 3.3. Moreover, withdrawal is not viable in either vignette without losing operational capabilities, given that this is a fixed installation, and the narrative is far from stable: It could readily be manipulated to present a very negative view of U.S. forces and their actions.

We also explored other potential interactions in which NLWs might be used, although these did not ultimately take the form of vignettes. For example, a recurring problem around some U.S. bases overseas is local individuals attempting to steal fencing materials or other items along the base perimeter at night. These individuals are not always driven off by searchlights alone, and they may abscond before security forces can reach them. This problem can contribute to gaps in physical security while also conditioning security personnel to be less reactive to movements along the perimeter, contributing to complacency. Hostile forces can also query or co-opt petty thieves to garner information on the facility and its security responses. Such NLWs as laser dazzlers or the ADS could potentially help

TABLE 3.3
Vignette Summary

Number	Vignette	Domain	Service	GCC	Escalatory Adversary	Withdrawal Possible	Stable Narrative
1	Joint Surveillance and Target Attack Radar System (JSTARS) intercepted by People's Liberation Army Navy and Air Force (PLANAF) fighters	Air	USAF	INDOPACOM	X	X	X
2	Motorized confrontation with Russian forces in Syria	Ground	Army	CENTCOM	X	X	
3	Boats approach a destroyer in the Strait of Gibraltar	Maritime	USN	EUCOM	X		X
4	Ground threat around domestic humanitarian assistance/disaster relief (HA/DR) sites	Ground	Army, USMC, National Guard	NORTHCOM	X		X
5	Maritime threat around domestic HA/DR sites	Maritime	USN, USCG	NORTHCOM	X		X
6	Securing embassy in Bahrain	Ground	USMC	CENTCOM	X		
7	Demonstrators in Palau	Ground	USAF	INDOPACOM	X		
8	Lasing in Palau	Ground	USAF	INDOPACOM	X		

Table 3.3—Continued

Number	Vignette	Domain	Service	GCC	Escalatory Adversary	Withdrawal Possible	Stable Narrative
9	Maritime standoff in the South China Sea	Maritime	USN, USCG	INDOPACOM		X	X
10	Maritime standoff in the Arctic	Maritime	USCG	EUCOM		X	X
11	Blockade enforcement near Venezuela	Maritime	USN	SOUTHCOM		X	X
12	SOF hostage rescue in Somalia	Ground	SOF	AFRICOM		X	
13	Expeditionary Advanced Base (EAB) defense against UAVs in the Philippines	Multi	USMC	INDOPACOM			

NOTES: A cell with an X means the vignette has that feature; an empty cell means the vignette does not have that feature. AFRICOM = U.S. Africa Command; EUCOM = U.S. European Command; INDOPACOM = U.S. Indo-Pacific Command; NORTHCOM = U.S. Northern Command; SOF = special operations forces; SOUTHCOM = U.S. Southern Command; USAF = U.S. Air Force; USCG = U.S. Coast Guard; USMC = U.S. Marine Corps; USN = U.S. Navy.

25

prevent recurrent intrusions. However, because this is a chronic problem rather than an acute encounter, we did not transform it into a vignette. Similarly, we considered the idea of using NLWs in response to surprise attacks on a base, where the speed-of-light or speed-of-sound effects of some NLWs could be advantageous while also reducing the risk of collateral damage or fratricide when there was uncertainty about the location of the attackers. However, despite repeated attempts, we did not identify a vignette in which NLWs would not be superseded by the use of exclusively lethal weapons in this context.

We also recognized that there are many possible vignettes that could be explored. For example, we had considered NLW usage as part of military response to a pandemic event, such as U.S. military HA/DR in West Africa following the 2014 Ebola outbreak. However, we assessed that the dynamics of using NLWs for security in a pandemic would not be profoundly different from a non-pandemic situation: Regardless of disease considerations, potentially hostile individuals should be kept at a distance well beyond the range at which they could possibly infect U.S. personnel. We also considered a vignette describing HA/DR overseas, because DoD has performed that mission numerous times around the globe in recent decades. However, this would have many similarities with the domestic HA/DR situations that we explored, so we chose not to develop additional vignettes in this vein. Vignettes involving pandemics, overseas HA/DR, both situations, or many other subjects could be explored in subsequent analysis.

Evaluating Metrics in the Context of Vignettes

Once we developed our set of vignettes, we explored them using the logic model and associated metrics described above and in the previous chapter. The research team ascertained what types of NLWs would potentially be effective in the context of each vignette, based on what U.S. forces were trying to achieve, the overall context (including what populations are present; the physical environment; the distances involved; and escalatory, withdrawal, and narrative considerations), and potential risks. These items are listed in Table 3.4. To continue the extended discussion of vignette 7 from earlier, in which demonstrators were blocking base gates in Palau, the use of acoustic hailers can underscore the seriousness of demands to disperse; this will induce some individuals to leave and clarify the more serious determination of others. Briefly and periodically discomfiting particular individuals or sections of the crowd with an ADS could help disperse some of those who remain. If there are still holdouts, pepper balls could cause them

TABLE 3.4

NLW Usage in Vignettes

Vignette	NLWs Used	Reasons for Selecting Specific NLWs
1. JSTARS intercepted by PLANAF fighters	LROI	Intermittent LROI use provides a rapid, effective way of demonstrating resolve to the other pilots and makes it harder for them to maneuver without dangerously incapacitating them
2. Motorized confrontation with Russian forces in Syria	AHD, LROI, EoF CROWS, PEVS, RFVS	Various hailing and dazzling NLWs can be used to communicate and make it harder to drive, while the PEVS and RFVS can prevent their vehicles from approaching
3. Boats approach a destroyer in the Strait of Gibraltar	AHD, LROI, VIPER, MVSOT	Hailing and dazzling can communicate and help differentiate intent while also making it harder to drive a boat in a particular direction; VIPER and MVSOT can prevent boats from approaching
4. Ground threat around domestic HA/DR sites	ADS, AHD, EoF CROWS, LROI, OI	Hailing and dazzling can communicate and help differentiate intent while an ADS can incapacitate individuals and induce them (or others) to depart
5. Maritime threat around domestic HA/DR sites	ADS, AHD, EoF CROWS, LROI, MVSOT, pepper balls, beanbag rounds, rubber bullets, VIPER	Hailing and dazzling can communicate and differentiate intent, ADS and pepper balls can incapacitate individuals and/or induce departure, beanbag rounds and rubber bullets inflict limited injuries to incapacitate or drive away particularly recalcitrant individuals, and VIPER and MVSOT prevent boats from approaching
6. Securing embassy in Bahrain	ADS, AHD, pepper balls, beanbag rounds, rubber bullets	Hailing can communicate and differentiate intent, ADS and pepper balls can incapacitate individuals and/or induce departure, and beanbag rounds and rubber bullets inflict limited injuries to incapacitate or drive away particularly recalcitrant individuals
7. Demonstrators in Palau	ADS, AHD, pepper balls	Hailing can communicate and differentiate intent by causing some individuals to disperse, ADS and pepper balls can incapacitate individuals and/or induce departure

Table 3.4—Continued

Vignette	NLWs Used	Reasons for Selecting Specific NLWs
8. Lasing in Palau	ADS, AHD, EoF CROWS, LROI, pepper balls	Hailing can cause some individuals to drop their dazzlers and flee, while dazzling them, using ADS, or pepper balls can do that or incapacitate them so that they are no longer able to dazzle pilots
9. Maritime standoff in the South China Sea	ADS, LROI	Temporarily dazzling personnel driving the ship and discomfiting personnel on deck can demonstrate resolve and cause them to want to back away; hailing is unnecessary, because other communications channels exist
10. Maritime standoff in the Arctic	ADS, LROI	Temporarily dazzling personnel driving the ship and discomfiting personnel on deck can demonstrate resolve and cause them to want to back away; hailing is unnecessary, because other communications channels exist
11. Blockade enforcement near Venezuela	ADS, LROI	Temporarily dazzling personnel driving the ship and discomfiting personnel on deck can demonstrate resolve and cause them to want to back away; hailing is unnecessary, because other communications channels exist
12. SOF hostage rescue in Somalia	Beanbag round, rubber bullets, flash-bang grenades	These munitions can temporarily incapacitate personnel or inflict light injuries while limiting the risk of accidentally killing hostages
13. EAB defense against UAVs in the Philippines	C-UAV	Disabling the UAV without shooting it demonstrates resolve and impedes its mission while managing escalation

NOTES: Acoustic: The AHD communicates orally at long distances.

Laser dazzlers: The OI and LROI create intense glare without permanent effects.

Escalation of Force Common Remotely Operated Weapons Station (EoF CROWS): This combines both acoustic and laser-dazzling capabilities.

ADS: This emits millimeter-wave energy to create a temporary heating sensation.

Electronic disruption: The PEVS, RFVS, VIPER, and C-UAV systems emit microwave energy to disrupt electronics.

Mechanical disruption: MVSOT halt or slow the movements of small vessels by entangling, coating, or otherwise affecting their propellers.

Munitions: Beanbag rounds, rubber bullets, and flash-bang grenades deliver blunt-impact effects and/or temporary incapacitation with bursts of light or sound.

Riot control agents: Pepper balls can be launched to disperse irritating pepper spray over a wide area.

to flee, or at least make it easy to take them into custody, given a degree of incapacitation. In vignette 8, involving the adolescents who were targeting pilots with laser pointers, acoustic hailing might frighten them into stopping what they are doing and fleeing. The use of laser dazzlers might compound this effect and cause the adolescents to be unable to accurately aim their laser pointers at moving targets. If they still attempt to persist, an ADS and pepper balls will further reduce their ability to aim accurately while also causing them to flee and perhaps be sufficiently deterred not to try this again. More details on these and other vignettes appear in Appendix C.

Acoustic Systems, ADS, and Laser Dazzlers Were Particularly Versatile NLWs

Table 3.5 summarizes how frequently we used each of the classes of NLWs across all vignettes. Acoustic systems, the ADS, and laser dazzlers were used in a slight majority of the vignettes across multiple domains. Although versatility is far from the only driver of prioritization, and the set of vignettes was not designed to precisely reflect the frequency or relative importance of different situations that U.S. forces encounter, these relative numbers indicate that acoustic systems, ADS, and laser dazzlers can be applied in a wide variety of contexts. To the extent that JIFCO is seeking to focus its energies on programs that address overall DoD needs, these classes of NLWs can be

TABLE 3.5

Frequency of NLW Usage Across 13 Vignettes

NLW Type	Number of Vignettes Used
Laser dazzler	9
ADS	8
Acoustic	7
Munition	4
Riot control agent	4
Electronic disruption	4
Mechanical disruption	2

emphasized. Other types of systems, however, are also important in specific contexts and should continue to be pursued.

Evaluation of Metrics

For each vignette, we collectively assessed which elements of the logic model, and which metrics that related to them, were applicable. (The others were designated as N/A.) We then collectively assessed the validity, reliability, feasibility, and timeliness of each applicable metric in the context of the vignette, using a three-point scale (low, medium, high) for each criterion. We made these assessments based on the scales provided in Table 3.6, and we will demonstrate an example of such an assessment after the table. We recognize that these assessments are themselves somewhat subjective, and there may be instances where others would ascertain different values; however, it is unlikely that assessments by others would be radically different,

TABLE 3.6

Scales for Evaluating Validity, Reliability, Feasibility, and Timeliness

	Validity	Reliability	Feasibility	Timeliness[a]
High	Directly measures the element or a close proxy	Well-defined, objective, and stable	Required data sets are readily available and user-friendly	Hours
Medium	Closely related to the element being measured	Some ambiguity, subjectivity, and/or volatility	Required data sets could be collected with limited effort	Days
Low	Indirectly related to the element being measured	Considerable ambiguity, subjectivity, and/or volatility	Required data sets would be challenging to collect	Weeks to years

SOURCE: Savitz, Matthews, and Weilant, 2017.

[a] Note that this criterion refers to the timeliness of receipt of the values of metrics, not timeliness of the effects of NLWs. We selected the values for high, medium, and low timeliness as follows. Values of metrics that are received within hours can inform short-term tactical decisions. Those that are available within days may affect larger operational activities. Those that take weeks or longer can inform future operations.

and the process we used is both explicit and traceable, helping to make discussion of particular points more concrete.

For illustrative purposes, here is an example. One of the outputs is "avoiding alienation of the population, host-nation forces, and the host government." One of the metrics used to measure it is "host nation public opinion on use of NLWs, measured by polls." In the vignette in which there are demonstrators outside a base in Palau, this metric scores high for validity: It measures how well the host nation is perceiving NLWs, which is a close proxy for avoiding alienation. It scores medium for reliability: There is always some uncertainty in polling, which may be exacerbated by such factors as a small sample size and the way in which the poll is conducted (e.g., if some demographics are more represented than others in the sample). Polling takes considerable time and effort, so the metric scores low on both of those counts. The values associated with a metric can vary among vignettes, or a given metric may not be relevant at all in a different vignette. For example, the same metric would be N/A in the context of the Arctic standoff, because there is no host nation in that vignette.

In all, we made 5,980 qualitative assessments across 115 metrics, four rating types, and 13 vignettes. Appendix B provides each metric, the number of vignettes in which it was applicable, and the average value of its validity, reliability, feasibility, and timeliness across all of the vignettes for which it was applicable.

Using the statistical analyses of the qualitative assessments, two key themes emerged. The first was that the metrics generally scored well across the vignettes in which they were applicable, with the exception that outcome metrics had more limited feasibility and timeliness, as might be expected, given the greater difficulty of measuring higher-level elements of the logic model. The second was that a large fraction of the metrics applied to only some of the vignettes, which underscores the need to tailor the selection of metrics to the particular context in which they are used. We discuss these findings next.

Metrics Generally Score Well, Except Outcome Feasibility and Timeliness

To transform the high, medium, and low values into quantitative ones that could be averaged, we set a high score equal to 10, a medium one equal to 5, and a low one equal to 0. The average scores for activity, output, and outcome metrics across all vignettes in which those metrics were applicable are shown in Figure 3.1.

Clearly, metrics across all three categories scored relatively well in terms of validity. In other words, they generally did a good job of measuring the element of the logic model that they were intended to measure. They did a little less well in terms of reliability, reflecting the fact that some measurements may be uncertain—for example, the percentage of a targeted population experiencing NLW effects may not be clear. Feasibility and timeliness scores were high for activity and output metrics but substantially lower for outcome metrics. This reflects the difficulty of ascertaining the values of some of the outcome metrics, as well as the long time lags before the full effects of NLW usage emerge at the outcome level. For example, public perceptions of NLW usage take time to emerge and require substantial resources to evaluate. Similarly,

FIGURE 3.1

Average Scores for Activity, Output, and Outcome Metrics Across all Vignettes in Which They Were Applicable

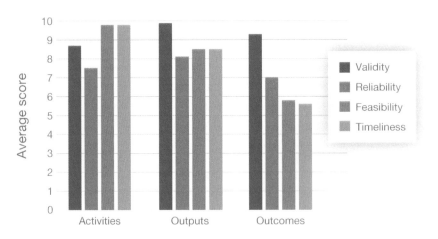

NOTE: Averaging based on high = 10, medium = 5, low = 0.

although NLW usage could contribute to enhanced morale due to reduced risk of lethal collateral damage, this might take years to become clear, then require copious data-gathering and analysis to corroborate.

One key issue that was not captured in the above assessment is that the values of some of the metrics depend on repeated trials—for example, the percentages of cases in which NLWs enabled pre-emptive action, curtailed the other side's options, or affected the other side's behavior. These are useful if tactical situations are repeated but less relevant in the context of singular, unique events.

However, Metrics Do Not Apply to All Vignettes

One significant caveat to our findings is that some elements of the logic model may be applicable to some vignettes but not others. Of the 5,980 assessments we made, 2,795 (47 percent) were judged to be not applicable (N/A) to a logic model element, which is not accounted for in the above analyses. For example, in vignette 7, involving demonstrators obstructing access to an air base in Palau, metrics related to the activity of incapacitating infrastructure or materiel, the output of gathering intelligence from captured personnel and materiel, or the outcome of avoiding rebuilding costs are irrelevant.

A detailed breakdown of how prevalent N/As were across vignettes and classes of logic model elements is shown in Table 3.7. In general, activity and output metrics were mostly applicable across the broad range of vignettes; only four of the 26 observations exceeded 40 percent, and only one exceeded 50 percent. On the other hand, for five of the 13 vignettes, more than half of the outcome metrics were inapplicable. This reflects the specificity of many of the outcome elements: Those that relate to such subjects as demonstrating resolve in the context of competition, partner cooperation, and public perceptions only relate meaningfully to only some vignettes.

The data in Table 3.7 highlight the need to tailor the logic model to meet specific contexts, particularly with respect to outcomes. In evaluating NLW usage in a particular context, only a subset of the metrics the RAND team identified should be measured.

TABLE 3.7

Percentage of Inapplicable Metrics by Class of Logic Model Element and Vignette

Vignette	Activity (%)	Output (%)	Outcome (%)
1. JSTARS intercepted by PLANAF fighters	28	33	39
2. Motorized confrontation with Russian forces in Syria	27	24	49
3. Boats approach a destroyer in the Strait of Gibraltar	20	25	55
4. Ground threat around domestic HA/DR sites	18	25	58
5. Maritime threat around domestic HA/DR sites	10	14	76
6. Securing embassy in Bahrain	16	21	63
7. Demonstrators in Palau	27	47	27
8. Lasing in Palau	34	49	17
9. Maritime standoff in the South China Sea	29	20	51
10. Maritime standoff in the Arctic	33	36	32
11. Blockade enforcement near Venezuela	25	37	37
12. SOF hostage rescue in Somalia	51	27	22
13. EAB defense against UAVs in the Philippines	42	35	23
Percentage across all vignettes	29	31	39

Closing Remarks

In this chapter, we have described how the team identified metrics to evaluate the impact of NLWs, by measuring elements of the logic model (activities, outputs, and outcomes) that relate to the tactical, operational, and strategic levels. The evaluations of these metrics, which are further elucidated in Appendix B, reveal their overall utility and the relative merits of individual metrics for different types of situations. Using these metrics and their characterization, JIFCO can work with the services to collect data to assess the values of specific metrics, which can then provide specific insights on the impact of NLWs at multiple levels.

CHAPTER FOUR

Themes Identified in Interviews

Much of what we learned about NLWs was gained by interviewing technologists involved in their development, policy-related personnel who provide resources and govern their use, and the users who ultimately use NLWs in operations. In this chapter, we examine their perspectives directly, which complemented some of the insights that we garnered in the process of developing the logic model and metrics, then evaluating the metrics in the context of vignettes.

We will first describe our methodology and introduce the different thematic categories that we saw in our interviews. We will then explore four overarching observations that emerged from our analysis. The full interview protocol and codebook is available in Appendixes D and E, respectively.

Approach

Our interview analysis approach had three parts: recruitment, data-gathering, and analysis.

Recruitment and Preparation

We recruited interviewees with the assistance of our project sponsor. We included personnel with three different perspectives:

1. **Users:** Individuals who were responsible for employing NLWs. We spoke primarily with commanders and staff officers who are responsible for planning NLWs' tactical usage and ensuring that they are employed correctly; some also had direct experience of employing NLWs themselves.

2. **Policy experts:** Individuals who are responsible for developing NLW doctrine, TTPs for their use, training, and resourcing NLW programs.
3. **Technologists:** Individuals who develop the NLWs themselves.

We conducted 36 interviews of both individuals and groups, across 25 distinct organizations. Eight of these 25 organizations included NLW users with operational experience across multiple domains and geographic areas. Twelve organizations were policy offices from across the services as well as JIFCO and OSD. Five were technology-centric organizations from various service, OSD, and contractor centers.

Data-Gathering

We used a semistructured interview approach that gave interviewees latitude to discuss a wide range of NLW issues. We developed and refined a protocol internally and conducted interviews telephonically because of COVID travel restrictions. We conducted interviews with at least two researchers present; one led the conversation, and the other took detailed written notes that we then used for our analysis.

Data Analysis

Our interviews yielded a large volume of rich qualitative data. To systemically analyze it, we used a thematic analysis approach, enabled by Dedoose analytic software.[1] This is an iterative process in which readers identify (or code) recurring patterns in the interview notes (or *themes*) and relationships among them.[2] The themes and subthemes were initially developed

---

[1] Dedoose, version 8.0.35, web application for managing, analyzing, and presenting qualitative and mixed method research data, Los Angeles: SocioCultural Research Consultants, 2018.

[2] For more on the practice of thematic analysis in the social sciences, see Greg Guest, Kathleen M. MacQueen, and Emily E. Namey, *Applied Thematic Analysis*, Thousand Oaks, Calif.: Sage Publications, 2011; and Eli Lieber, "Mixing Qualitative and Quantitative Methods: Insights into Design and Analysis Issues," *Journal of Ethnographic and Qualitative Research*, Vol. 3, No. 4, 2009, pp. 218–227.

deductively, based on the overall research aims, reviewing literature, and initial notes and impressions after stakeholder interviews. They were then refined through inductive reading and re-reading of an initial random sample of interview notes, followed by refinements as the coders read the entire body of interview notes. The result is a set of themes and subthemes that balances the nuances in the data with a model from which more generalizable observations can be characterized. The hierarchy of codes allows us to explore the data at different levels of abstraction (see Table 4.1).

Once the themes and subthemes were identified, coders read all interview notes and highlighted statements (or excerpts) that reflected the rel-

TABLE 4.1
NLW Interview Themes

Theme	Subtheme (if applicable)	Definition
Challenges	Cultural	Challenges or potentially negative issues related to organizational or societal norms that affect how NLWs are used when it is technically and operationally possible within policy (e.g., commander does not think to use an NLW capability when it is possible)
	Operational	Challenges or potentially negative issues that relate to how NLWs are used (i.e., the TTPs used by NLW operators) in accordance with established policies and technical capabilities
	Policy	Challenges or potentially negative issues that relate to regulations, policies, or laws governing when or how NLWs should or should not be used (excludes challenges regarding how NLWs are used in practice, which is covered in operational challenges)
	Resources	Challenges or potentially negative issues to broader NLW usage related to availability of resources to develop, field, or train on NLWs
	Technical	Challenges or potentially negative issues to using NLWs related to physical characteristics (such as form factor) or effect characteristics (such as range)
	Other	Any challenge not defined by any others in this list of themes
Evaluation		Excerpt related to metrics evaluating the utility of an NLW

Table 4.1—Continued

Theme	Subtheme (if applicable)	Definition
IFC definition		Excerpt provides a characterization of what capabilities should be considered IFCs. For example, a discussion on whether or not cyber tools count as IFCs. Excludes discussions on cultural place of IFCs within an organization.
Opportunities		Excerpt identifies an advantage or opportunity to use an NLW in an operational setting (excludes discussions on opportunities to advance the business case for NLWs within an organization)
Scenario described	Non-great-power conflict	Use case describes traditional crowd control or entry control point use during counterinsurgency, counterterrorism, peacekeeping/peace enforcement, and support to civil authority settings.
	Great-power conflict	Use case is related to great-power conflict: It includes usage during conventional deterrence, competition phase activities against a peer adversary, or use in offensive or defensive missions against a peer adversary.

SOURCE: RAND analysis of interview notes.

evant themes. Interviews were read by one of two coders. Both coders were trained and tested to ensure that they followed a consistent definition of each code. A test of sample codes was developed in Dedoose; the Pooled Cohen's Kappa (a value representing the level of agreement between coders) was calculated to be 0.69, indicating good agreement.[3] We also assigned a perspective to each interviewee. This enabled us to characterize different themes based on user, technologist, and policy perspectives.

Emergent Insights from Interview Analysis

The themes and insights that emerged from our interview efforts highlighted numerous challenges to greater NLW adoption; 56 percent of all

---

[3] Matthew B. Miles and Michael A. Huberman, *Qualitative Data Analysis: An Expanded Sourcebook*, 2nd ed., Thousand Oaks, Calif.: Sage Publications, 1994.

interview excerpts coded were related to one or more NLW challenges, despite the interview protocol devoting only two of its ten questions to the topic. Other issues were discussed at times, but not nearly to the degree that challenges were considered. When describing challenges, interviewees articulated issues that defy simple characterizations. Interviewees described various issues that intertwined and interacted with each other. Figure 4.1 summarizes the frequency of each.

The following key themes emerged from our analysis:

- Cultural and resource issues are the greatest challenges to NLW adoption. Cultural issues primarily related to a reticence to embrace NLWs even when doctrine and policy allowed for their use. This reticence often related to potential users having little confidence in NLWs working as intended, not seeing them as useful compared with more-lethal capabilities, or not fully understanding the effects of NLWs. In terms of resource challenges, interviewees highlighted that a lack of NLW availability and competing training demands often forced them to de-emphasize NLWs even when they might have been useful.

FIGURE 4.1

Challenges Mentioned by Interviewees

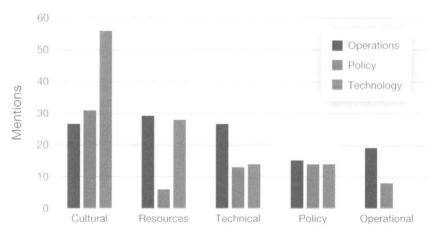

NOTE: The figures are normalized to account for the different number of interviews for each type conducted.

- NLWs are often perceived as burdensome to the point that they are not carried into operational engagements due to logistical concerns and constraints.
- Challenges interact and reinforce each other. An example of this is that commands with little familiarity with NLWs tend to discount their utility, so they limit the extent of NLW training and usage, which reinforces that unfamiliarity.
- Opportunities for additional NLW usage are not widely recognized. For example, interviewees generally had little to say about the potential applicability of NLWs in strategic or great-power competition beyond limited perception of NLW usage in gray-zone situations.

We further discuss these themes next.

Cultural and Resource Issues Are the Greatest Challenges to NLW Adoption

Cultural issues related to a reticence to embrace NLWs even when doctrine and policy allow their use was the most common challenge cited. Some interviewees appear to have little confidence that NLWs will work as intended, or did not see them as useful compared with more-lethal weapons. More deeply though, all groups implied that they did not fully understand NLWs' physical effects; many stated that they were worried about using them in a way that would invite subsequent punishment (e.g., concerns about the intensity or permanence of ADS effects, and the fear of being punished for using it were mentioned often). This suggests that DoD is struggling to make sense of NLW effects and how to leverage them even in situations in which DoD has had significant experience using them, such as crowd-control situations.

Resource challenges were also mentioned frequently by technologists and users. Users highlighted a lack of NLW availability and competing training demands that often forced them to de-emphasize NLWs even when they might have been useful. Technologists mentioned wanting more material resources to develop and institutionalize NLWs.

Other challenges were mentioned, but with less frequency. Some commented on technical limitations of NLWs, noting that they do not provide truly unique effects. One interviewee stated,

> If [NLWs] can produce something that is incredibly effective or does something new, it may be a gamechanger. Otherwise, it won't really work for us at the moment because it doesn't seem that there [is] any itch needing to be scratched.

Surprisingly, there was little discussion about policy challenges of greater NLW adoption. Some interviewees expressed concerns about inconsistent NLW use policy across parts of DoD, but these perspectives were from personnel who use them in law enforcement roles.

> But if I'm at a base where I carry a taser but not pepper spray, and taser has manufacturer-specific training, I may go to a different base and have a different set of systems and training.

This last point speaks to the desirability of standardizing NLW policies, CONOPS, TTPs, and training when possible.

NLWs Are Often Viewed Negatively, Within DoD and Beyond

Part of the cultural issue reflects a stigma associated with NLWs, even among DoD personnel, that was found based on both interviews and the literature review. Perceptions of NLWs among the general public are also often negative, reinforced by news media sometimes sensationalizing their effects, regardless of demonstrations of their absolute and relative safety.[4] As one interviewee stated:

> There is a psychological barrier to overcome here—somehow it is more acceptable to kill [or] injure people than to annoy them.

In addition, all of the services focus heavily on the use of lethal force; this was recently reinforced by the 2018 National Defense Strategy's stated goal of enhancing lethality.[5] In this context, the concept of *non-lethal weapons* has reduced traction, which was one of the motivating factors behind that

---

[4] See, for example, Michael D. Shear, "Border Officials Weighed Deploying Migrant 'Heat Ray' Ahead of Midterms," *New York Times*, August 26, 2020. For information on the safety record of the ADS described in that document, see LeVine, 2009.

[5] Mattis, 2018.

phrase being replaced with *intermediate force capabilities* in 2020: They provide additional options along a continuum.[6]

NLWs Are Often Perceived as Burdensome

Users often viewed NLWs, given their current dimensions, as forcing them to make a binary choice between carrying or mounting them and having lethal weapons available in their place. One interviewee spoke about choosing to carry NLWs as a form of risk:

> If I'm commanding a small ship . . . the previous version of the laser dazzler went from something handheld to something needing a tripod. So they were putting it on boats with two .50 caliber gun mounts and saying, "You can either have a gun or this laser." I personally wouldn't pick the laser. Options are important to control escalation, but if it's an either/or choice, I'll default to defending my ship.

Another interviewee stated a similar sentiment more bluntly:

> It is always problematic when you are trying to shove ten pounds of [expletive] into a five pound bag. Form factor and weight are critical and carrying ten pounds of NLWs equates to leaving 10 pounds of [operational load] (i.e., ammunition) or just carrying more, that is the cultural problem you are going to run into. The further you move from a wide-area security mission to a mission with the potential for lethality, the more likely you are to see dumping of that gear, at large form factors.

These considerations, and the fact that many situations might call for either NLWs or the use of lethal force, speak to the desirability of designing NLWs to minimize space, weight, and power requirements, even at the expense of other attributes, such as range. Systems need to be compact enough to avoid being put aside at some point along the logistical chain.

---

[6] Wendell B. Leimbach, Jr., "DoD Intermediate Force Capabilities: Bringing the Fight to the Gray Zone," Joint Intermediate Force Capabilities Office, U.S. Department of Defense, Non-Lethal Weapons Program, March 2020, Distribution A: Approved for Public Release.

Challenges Interact and Reinforce Each Other

Interestingly, interviewees often discussed how challenges reinforce each other. Resource limits prevent them from getting more familiarity with NLWs, which keep them from identifying policy gaps. Lack of widely understood TTPs in some communities discourage commanders and their staffs from seeking out NLW resources. One user who was exposed to NLWs during a deployment articulated one dimension of this problem:

> Because we're not training with it, and you only get exposed to it in country [or] right before you arrive, or there wasn't a repair plan and equipment was faulty, so you ended up not using it. Weren't coming up with novel ways to use it. Not just [NLWs], got a lot of equipment over that time that you weren't trained in using, so didn't really know how to use it so it got left in the depot. You need to build it into the training if you are going to utilize the equipment.

When asked about why his unit did not use NLWs even though they were available, one interviewee responded bluntly:

> [It was] a conscious choice by commanders to not bother. NLWs are just not worth it to them. The only current gap [worth mentioning] is the laser dazzler. [We] need a more effective dazzler that can operate in daylight.

These responses suggest that a lack of familiarity with NLWs ultimately creates a cultural barrier to their use; operators are not acquainted with them and are not trained on any form of employment, so they discount their utility, which reinforces the lack of familiarity. This point reinforces the need for (1) training and familiarization prior to deployment and (2) integration into standard TTPs.

Opportunities for Additional NLW Usage Are Not Widely Recognized

Interviewees, particularly users, could clearly articulate NLW uses from prior operational experiences in the context of crowd control, vehicle checkpoints, law enforcement, and fixed-site security. However, interviewees did not perceive clear NLW use cases in great-power competition, beyond generally mentioning that they might be relevant in gray-zone scenarios. Even interviewer prompts did not motivate further responses regarding how they could be used. Moreover, interviewer inquiries about whether NLWs could be used to complement lethal force when an attacker was already using lethal force were sharply dismissed, on the grounds that lethal force was the only appropriate response to lethal force.

The fact that NLWs were not widely perceived as having potential roles beyond the traditional ones, such as law enforcement and crowd control, suggests that there is a need to articulate and demonstrate these possibilities more widely in doctrine, policy, plans, wargames, exercises, and training.

Concluding Remarks

In this chapter, we have described key insights gleaned from the interviews, in addition to those that directly contributed to the development of the logic model and metrics. A lack of confidence in and understanding of NLWs, coupled with perceptions of them as somehow more damaging than lethal weapons, also have implications for the logic model and associated metrics. When JIFCO and other NLW stakeholders are engaging with other parts of DoD and the wider public, they can use the logic model to clearly articulate the activities that NLWs perform. To the degree that the effects of these weapons are clarified and made concrete, they are less likely to be perceived as vaguely sinister. Accurate perceptions by service members, in turn, can contribute to greater resource allocation so that adequate training, logistics, and maintenance support are provided to ensure that these systems can be used effectively when needed. Emphasizing the fact that these systems contribute to strategic goals enumerated in the National Defense Strategy can also crystallize understanding of how they relate to wider military aims.

The fact that these systems are often perceived as burdensome underscores the value of collecting data on the degree to which they actually impinge on lethal capabilities. By assessing the values of metrics related to the output of tactical resource costs—for example, cost differentials between NLWs and lethal systems, capacity reductions for lethal systems, NLWs' logistical burdens, and NLWs' support requirements—JIFCO and other stakeholders can ascertain the extent to which these perceptions are well-founded in different contexts and find ways in which to address shortfalls. Metrics related to the output of reducing adversary options and imposing costs may also be useful in this context; they help indicate the extent to which whatever perceived burdens NLWs do impose are offset by the burdens they impose on adversaries.

CHAPTER FIVE

Conclusions and Recommendations

This report has described how the impact of NLWs can be measured more effectively. Much of this analysis was structured around the development of a logic model that linked the NLWs and their context with the activities that they perform, the direct outputs of those activities, higher-level outcomes, and DoD-wide strategic goals. Drawing on that structure, the team identified a series of metrics that could be used to measure the elements of that logic model. Furthermore, the study team characterized those metrics in terms of how well they measured each element in the context of varied vignettes and how consistently, easily, and quickly those measurements were made. This lays the groundwork for data collection to assess the values of those metrics in ways that enable evaluation of the tactical, operational, and strategic impact of NLWs.

Conclusions

Results from the Logic Model and Evaluated Metrics

All seven of the activities in the logic model ultimately have strong connections to the strategic goals, as do nine of the 13 outputs and five of the nine outcomes. These elements (highlighted within bold blue boxes in Figure 2.3) and their associated metrics can be used to make the strongest case for the strategic impact of NLWs at a DoD-wide level.

Second, when we examined the metrics developed for the elements in each level of the logic model, a few patterns emerged. Activity metrics primarily relate to which people or systems are affected by NLW usage and how well they respond to NLWs. Output metrics, by contrast, generally relate to providing the user with more time and options, curtailing the adversary's options, and reducing tactical risks. Finally, outcome metrics most often relate to reducing strategic and operational risks, influencing perceptions, maintaining morale, and reducing costs.

When applying the logic model and metrics to the vignettes, our analysis also revealed which NLWs were generally the most applicable to the range of contexts encompassed by our vignettes. We found that the IFCs that were particularly versatile were acoustic systems and laser dazzlers used to hail, deceive, distract, disorient, or confuse and the ADS used to provide focused, discriminating effects that can tactically deter, deny access, or cause individuals to depart.

Combining this information could help JIFCO and other stakeholders structure discussions of how NLWs affect DoD's ability to achieve its tactical, operational, and strategic aims. For example, the direct tactical impact of NLW usage in a gray-zone encounter may be to affect another party's mobility: A ship's pilot, subjected to intense glare from a laser dazzler, chooses to divert the ship's course away from the confrontation. The operational impact is that the United States has demonstrated resolve while managing escalation. Meanwhile, the strategic impacts include helping to compete below the level of armed conflict and proactively expanding the competitive space.

Themes Identified in Interviews

Much of what we learned about NLWs was gained by interviewing 36 groups of experts and stakeholders across 25 organizations spanning three broad categories: technologists involved in NLW development, policy-related personnel who provide resources and govern NLW usage, and operators who ultimately employ NLWs. Four key themes emerged from our analysis:

1. **Cultural and resource issues are the greatest challenges to NLW adoption.** Cultural issues primarily related to a reticence to embrace NLWs even when doctrine and policy allowed their use. This reticence often related to potential users having little confidence in NLWs working as intended, not seeing them as useful compared with more-lethal capabilities, or not fully understanding the effects of NLWs. In terms of resource challenges, interviewees highlighted

that a lack of NLW availability and competing training demands often forced them to de-emphasize NLWs even when they might have been useful.

2. **NLWs are often perceived as burdensome** to the point that they are not carried into operational engagements due to logistical concerns and constraints.

3. **Challenges interact and reinforce each other.** An example of this is that commands with little familiarity with NLWs tend to discount their utility, so they limit the extent of NLW training and usage, which reinforces that unfamiliarity.

4. **Opportunities for additional NLW usage are not widely recognized.** For example, interviewees generally had little to say about the potential applicability of NLWs in strategic or great-power competition, beyond limited perception of NLW usage in gray-zone situations.

Recommendations

Leveraging the results of our analysis using the logic model, metrics, and series of vignettes, we recommend that JIFCO take a couple of natural next steps.

1. **Present and discuss the logic model in various forums, including with senior leaders, to convey how NLWs contribute to DoD strategic goals.** The logic model provides concrete descriptions of activities and relationships that have often been superficially or incompletely understood. As DoD continues to shift its focus toward competition with China and Russia, the logic model and exploratory vignettes make it clear how NLWs can contribute to that competition, including by explicitly linking NLWs to strategic goals from the National Defense Strategy, helping to counter some of the misperceptions and misunderstandings about NLWs that interviews revealed.

2. **Work with the services to collect data that can be used to evaluate the impact of NLWs by providing values for the metrics.** (Despite multiple searches, we were unable to unearth existing data sets that

provided the information that was needed.) The values of these metrics can be measured in real-world operations and potentially also in live exercises or wargames. Metrics that relate to outputs and outcomes that have strong links to strategic goals, are relevant to a range of vignettes, and are easy to measure should be prioritized. These metrics and their associated outputs and outcomes are shown in Table 5.1. The tables in Appendix B can reveal additional metrics that are relevant and easy to collect, among other attributes. Sets of metrics can also be tailored to reflect the types of tactical situations in which NLWs will be used. For example, the set of metrics to evaluate NLWs' impact in maritime standoffs with great powers will differ from the set used to evaluate NLWs' impact in dispersing a crowd of hostile individuals, or the sets used in other scenarios. Once the metrics have been selected for a particular context, their values can be measured in real-world operations, live exercises, and wargames.

Exploration of the 13 vignettes also demonstrated the utility of NLWs across a range of scenarios, beyond the uses in law enforcement and crowd-control to which they have often been relegated. For example, NLWs can play a role in pushing back during gray-zone confrontations with China or Russia, without risking dangerous levels of escalation due to the use of lethal force. However, the study also revealed a host of issues that inhibit the use of NLWs. Many of these relate to perceptions of NLWs as burdensome, lack of awareness regarding their prospective utility, lack of adequate unit-level training and integration into TTPs, and misunderstood or ambiguous policies. Negative perceptions of NLWs by different populations, including views that NLWs were more harmful than lethal weapons, also were cited by numerous experts. To overcome these factors, there are four main approaches that JIFCO should undertake:

1. Work with the services to ensure that policies and CONOPS are consistent and clearly understood.
2. Collaborate with the Joint Chiefs of Staff (JCS) J7 on joint training standardization regarding NLWs, to ensure that services provide thorough unit training with NLWs and that NLWs are tightly integrated into units' TTPs. Although the services direct their own

TABLE 5.1

Examples of Elements of the Logic Model and Associated Metrics

Outputs	
Element of Logic Model	Metric
Effectively responded to situations despite constraints	Percentage of tactical encounters in which use of NLWs was permissible, but lethal force was not
	Whether NLWs are allowed by ROE (binary yes/no distinction)
Reduced risk of U.S., partner personnel casualties	Percentage of tactical encounters with U.S. and/or partner casualties when NLWs were used relative to those when they were not
	Time required to switch from non-lethal to lethal capability if EoF is necessary
Element of Logic Model	Metric
Reduced adversary options and imposed costs	Change (absolute or percentage) in number of distinct options available to an adversary due to use of NLWs
	Percentage of encounters in which number of adversary options is reduced due to use of NLWs
	Percentage of encounters in which adversary experiences additional costs due to use of NLWs
Increased options for engaging targets	Change (absolute or percentage) in number of distinct options available due to use of NLWs
	Percentage of encounters in which number of options is increased due to use of NLWs
Outcomes	
Element of Logic Model	Metric
Competed effectively and demonstrated resolve while managing escalation in peacetime, gray-zone, and hybrid contexts	Percentage of incidents using NLWs that resulted in unwanted escalation divided by percentage of incidents not using NLWs that resulted in unwanted escalation
	Percentage of particular peacetime, gray-zone, and hybrid incidents in which NLWs were used
	Percentage of incidents in which NLWs were used and commanders perceived them as contributing effectively

Table 5.1—Continued

Outcomes	
Element of Logic Model	**Metric**
Avoided alienation of population, host-nation forces, and host government	Frequency and scale of protests and riots against U.S. presence, actions
	Frequency and scale of protests and riots related to events involving U.S. use of NLWs
	Frequency and scale of protests and riots against U.S. use of NLWs
	Degree of military cooperation/permissiveness (high, medium, low), as assessed by U.S. personnel
	Host nation forces' perception of U.S. use of NLWs, as assessed by U.S. personnel engaged with them
	Host government's perception of U.S. use of NLWs, as assessed by U.S. personnel engaged with them
	Frequency of negative public statements by government figures about U.S. use of NLWs
Element of Logic Model	**Metric**
Enhanced perceptions of U.S. forces (in United States and internationally)	Frequency and scale of protests and riots internationally against U.S. presence, actions in third country
	Frequency and scale of protests and riots in United States against U.S. presence, actions in another country
	Frequency and scale of U.S. protests and riots related to NLW use within the U.S. or along the border
Increased partner cooperation	Degree of military cooperation/permissiveness (high, medium, low), as assessed by U.S. personnel
	Number of joint exercises, patrols, or other activities between U.S. and partner nation forces

training, and JIFCO lacks authority in this area, the JCS J7 can help shape NLW training standards across DoD. JIFCO can also work directly with the services or particular units, given its interest in doing so, to ensure that units are adequately trained for NLW usage.

3. Shape perceptions within the military via explanations using the logic model, including exploration of vignettes, demonstrations both in live exercises and wargames, and the use of data sets to measure NLWs' impact, once those become available.

4. Future NLWs should be designed from the outset to minimize the aspects of them that contribute most to perceived and actual burdens. JIFCO should support the development of NLWs, as part of the Joint Capability Integration and Development System and Defense Acquisition System, to minimize the attributes that contribute most to perceived and actual burdens. Designing NLWs with a focus on ease of use, low maintenance, and reduced space, weight, and power requirements can make them more attractive to future users. Naturally, because design always involves a series of trade-offs, this might mean that those systems' capabilities are diminished in other respects, such as range. However, emphasizing certain features over others could make it more likely that these systems might be used on a larger scale. The study's finding that acoustic systems, an ADS, and laser dazzlers are particularly versatile can contribute to these systems being used in an array of contexts that might not previously have been fully realized. The fact that the ability of some NLWs to disable vehicles, vessels, or unmanned aircraft can be decisive in select contexts can also inform future usage.

Closing Remarks

This report describes how the tactical, operational, and strategic impact of NLWs can be characterized using a logic model and a set of associated metrics. This clarifies how these NLWs relate to DoD strategic goals, and, in tandem with observations from interviews about how NLWs are perceived, it facilitates better communication within DoD regarding how these systems can be better integrated into operations. The identification and character-ization of the metrics also lay the groundwork for data collection that can be used to further evaluate the impact of NLWs at multiple levels, which in turn can shape their usage in ways that enhance their contributions to DoD effectiveness.

Relationships Among Elements of the Logic Model

As discussed in Chapter Two, we characterized the strength of the relationships among elements of the logic model according to the following three-point scale:

- 2—strong, unequivocal connection
- 1—limited, indirect, or conditional connection
- 0—no connection.

We analyzed the strengths of relationships between each activity and each output, between each output and each outcome, and between each output and each strategic goal. This qualitative analysis is inherently somewhat subjective—a different group of analysts might have made select assessments differently—but it provides traceability and a basis for concrete discussion. Tables A.1 through A.3 show the relationships identified between elements in adjacent levels of the logic model, which are also portrayed graphically in Figures 2.3 and 2.4. The 0, 1, or 2 in each cell indicates the strength of the connection between the column and row headings for that cell.

In Table A.1, outputs are indicated by a letter, as follows:

Outputs

A. effectively responded to situations despite constraints
B. enabled pre-emptive action without appearing to be aggressor
C. increased options for engaging targets
D. reduced risk of exceeding ROE or Laws of War
E. reduced adversary options and imposed costs
F. gained time before deciding to take lethal action
G. enabled lower-signature clandestine operations
H. reduced risk of U.S., partner personnel casualties
I. minimized collateral damage and fratricide
J. reduced risk to U.S. systems or facilities
K. gathered intelligence from captured personnel and materiel
L. conserved and augmented lethal capabilities
M. reduced U.S. tactical costs (broadly defined)

TABLE A.1

Strength of Relationships Between Activities and Outputs

Activity	Output												
	A	B	C	D	E	F	G	H	I	J	K	L	M
Hail to clarify, demarcate, warn	1	2	1	2	0	2	0	1	1	1	0	0	1
Reveal other parties' intent	2	1	2	2	1	1	0	1	2	1	1	0	1
Deceive, distract, disorient, or confuse	2	2	2	2	0	2	1	2	1	2	0	2	1
Affect mobility: Slow, impede, halt, prevent from approaching or leaving, redirect, disperse, impel departure	2	2	2	2	2	2	1	2	2	2	1	2	1
Compel/ tactically deter: Convince others to take or not take specific actions	2	2	2	2	2	2	1	2	2	2	0	1	1

Table A.1—Continued

Activity						Output						
	A	B	C	D	E	F	G	H	I	J	K	L	M
Temporarily incapacitate personnel	2	2	2	2	2	2	2	2	2	2	2	2	1
Incapacitate infrastructure materiel	2	2	2	2	2	2	2	2	2	2	2	2	1

In Table A.2, outcomes are indicated by a letter, as follows:

Outcomes

A. competed effectively and demonstrated resolve while managing escalation in peacetime, gray-zone, and hybrid contexts
B. conducted operations in environments that were otherwise too dangerous due to collateral damage, fratricide, or escalation risks
C. avoided alienation of population, host-nation forces, and host government
D. enhanced perceptions of U.S. forces (in the United States and internationally)
E. increased partner cooperation
F. reused captured infrastructure and materiel
G. avoided rebuilding costs
H. set standards for partner nations
I. reduced negative effects on morale from collateral damage or substantially harming individuals without lethal intent

TABLE A.2

Strength of Relationships Between Outputs and Outcomes

Output	Outcome								
	A	B	C	D	E	F	G	H	I
Effectively responded to situations despite constraints	2	2	2	1	0	0	0	1	1
Enabled pre-emptive action without appearing to be aggressor	2	0	2	2	1	0	0	1	0
Increased options for engaging targets	2	1	1	1	0	1	1	0	1
Reduced risk of exceeding ROE or Laws of War	2	1	2	2	1	0	0	2	2

Table A.2—Continued

Output					Outcome				
	A	B	C	D	E	F	G	H	I
Reduced adversary options and imposed costs	2	0	0	0	0	0	0	0	1
Gained time before deciding to take lethal action	2	1	1	1	0	0	0	1	1
Enabled lower-signature clandestine operations	2	2	0	1	0	0	0	0	0
Reduced risk of U.S., partner personnel casualties	1	2	2	1	2	0	0	0	2
Minimized collateral damage and fratricide	1	2	2	2	1	2	2	2	2

Table A.2—Continued

	Outcome								
Output	**A**	**B**	**C**	**D**	**E**	**F**	**G**	**H**	**I**
Reduced risk to U.S. systems or facilities	1	0	0	0	0	0	0	0	1
Gathered intelligence from captured personnel and/or materiel	1	0	0	0	0	0	0	0	0
Conserved and augmented lethal capabilities	1	0	0	0	0	0	0	1	0
Reduced U.S. tactical costs (broadly defined)	1	0	0	0	0	0	0	0	0

TABLE A.3

Strength of Relationships Between Outcomes and Strategic Goals

Outcomes	Strategic Goals			
	Improve DoD's Competitive Advantage over Adversaries	Strengthen Alliances and Partnerships	Proactively Expand the Competitive Space	Improve DoD's Ability to Compete Below Level of Armed Conflict
Competed effectively and demonstrated resolve while managing escalation in peacetime, gray-zone, and hybrid contexts	2	2	2	2
Conducted operations in environments that were otherwise too dangerous due to collateral damage, fratricide, or escalation risks	2	2	2	2
Avoided alienation of population, host-nation forces, and host government	2	1	1	2
Enhanced perceptions of U.S. forces (in United States and internationally)	2	2	2	2
Increased partner cooperation	2	2	2	2
Reused captured infrastructure and materiel	0	0	0	0
Avoided rebuilding costs	0	0	0	0

Table A.3—Continued

Outcomes	Strategic Goals			
	Improve DoD's Competitive Advantage over Adversaries	Strengthen Alliances and Partnerships	Proactively Expand the Competitive Space	Improve DoD's Ability to Compete Below Level of Armed Conflict
Set standards for partner nations	0	1	0	0
Reduced negative effects on morale from collateral damage or substantially harming individuals without lethal intent	0	0	0	1

Metrics and Evaluations

In this appendix, we list the metrics associated with each of the elements of the logic model, the number of vignettes to which each metric was applicable, and the average rating for each metric in the context of the vignettes. The averages were calculated such that each *high* rating was given a value of 10, each *medium* rating was given a value of 5, and each *low* rating was given a value of 0. Instances where the metric was not applicable were not included in the calculation of the averages.

In reviewing these metrics, the reader should view each of them holistically, as many data points are relevant in considering the relative merits of different metrics. For example, a metric that is applicable to a wide range of vignettes may have its average score dragged down by one or two of them; this does not necessarily indicate that it is less useful than a less widely applicable metric that scores better. Although validity is the most important criterion, the others are also important—for example, garnering the value of a metric with low feasibility may not be possible if the resources to do so are limited. Also, the reader should not be surprised that many values gravitated toward the extremes of the scale, either zero or ten: This is a three-point scale with two of those points at the extremes, and there was often a degree of consistency of scores among the applicable vignettes, despite substantial differences among the vignettes themselves.

TABLE B.1
Activity Metrics

Activity	Metric	Number of Vignettes Applicable (out of 13)	Average Validity	Average Reliability	Average Feasibility	Average Timeliness
Hail to clarify, demarcate, and warn	Percentage of targeted population receiving communication	11	10.0	5.5	10.0	10.0
	Percentage of encounters in which non-targeted populations receive communication	4	0.0	3.8	7.5	7.5
	Percentage of targeted population responding as desired to receipt of communication	11	10.0	10.0	10.0	10.0
	Percentage of targeted population responding in undesired ways to communication	11	10.0	10.0	10.0	10.0
	Timeline between NLW use and response	11	9.5	10.0	10.0	10.0

Table B.1—Continued

Activity	Metric	Number of Vignettes Applicable (out of 13)	Average Validity	Average Reliability	Average Feasibility	Average Timeliness
Reveal other parties' intent	Percentage of targeted population experiencing NLW effects intended to reveal intent	5	9.0	5.0	9.0	10.0
	Percentage of encounters in which non-targeted populations are subjected to NLW effects	2	2.5	5.0	10.0	10.0
	Percentage of targeted population that responds in ways that reveal intent	5	10.0	9.0	10.0	10.0
	Percentage of targeted population that responds in ways that inaccurately suggest hostile intent (false positives)	5	10.0	5.0	10.0	9.0

Table B.1—Continued

Activity	Metric	Number of Vignettes Applicable (out of 13)	Average Validity	Average Reliability	Average Feasibility	Average Timeliness
Reveal other parties' intent (continued)	Percentage of targeted population that responds in ways that inaccurately suggest benign intent (false negatives)	4	10.0	6.3	10.0	8.8
	Timeline between NLW use and revelation of intent	5	10.0	10.0	10.0	10.0
Deceive, distract, disorient, or confuse	Percentage of targeted population experiencing NLW effects that are deceived, distracted, disoriented, or confused	8	10.0	5.6	10.0	10.0
	Percentage of encounters in which non-targeted populations are subjected to NLW effects	4	3.8	5.0	10.0	10.0

Table B.1—Continued

Activity	Metric	Number of Vignettes Applicable (out of 13)	Average Validity	Average Reliability	Average Feasibility	Average Timeliness
Deceive, distract, disorient, or confuse (continued)	Percentage of targeted population that responds in desired ways	8	10.0	8.8	10.0	10.0
	Percentage of targeted population that responds in undesired ways	8	10.0	8.8	10.0	10.0
	Timeline between NLW use and response	8	10.0	10.0	10.0	10.0
Affect mobility: Slow, impede, halt, prevent from approaching or leaving, redirect, disperse, impel departure	Percentage of targeted population experiencing effects that restrict mobility	11	10.0	6.4	9.5	10.0
	Percentage of encounters in which non-targeted populations are subjected to NLW effects	4	5.0	5.0	8.8	8.8

Table B.1—Continued

Activity	Metric	Number of Vignettes Applicable (out of 13)	Average Validity	Average Reliability	Average Feasibility	Average Timeliness
Affect mobility: Slow, impede, ... (continued)	Percentage of targeted population that responds in desired ways	11	10.0	9.1	10.0	10.0
	Percentage of targeted population that responds in undesired ways	11	10.0	9.1	10.0	10.0
	Timeline between NLW use and response	11	10.0	10.0	10.0	10.0
Compel/tactically deter: Convince others to take or not take specific actions	Percentage of targeted population experiencing effects of NLW	8	10.0	5.6	10.0	10.0
	Percentage of encounters in which non-targeted populations are subjected to NLW effects	2	2.5	5.0	10.0	10.0

Table B.1—Continued

Activity	Metric	Number of Vignettes Applicable (out of 13)	Average Validity	Average Reliability	Average Feasibility	Average Timeliness
Compel/tactically deter: . . . . (continued)	Percentage of targeted population that responds in desired ways	8	10.0	9.4	10.0	10.0
	Percentage of targeted population that responds in undesired ways	8	10.0	9.4	10.0	10.0
	Timeline between NLW use and response	8	10.0	10.0	10.0	10.0
Temporarily incapacitate personnel	Percentage of targeted population incapacitated by NLW	4	10.0	5.0	10.0	10.0
	Percentage of encounters in which non-targeted population is incapacitated by NLW	2	7.5	5.0	10.0	10.0
	Timeline between NLW use and incapacitation	4	10.0	10.0	10.0	10.0

Table B.1—Continued

Activity	Metric	Number of Vignettes Applicable (out of 13)	Average Validity	Average Reliability	Average Feasibility	Average Timeliness
Temporarily incapacitate personnel (continued)	Duration of incapacitation	4	10.0	8.8	10.0	10.0
Incapacitate infrastructure/ materiel	Percentage of targeted infrastructure/materiel incapacitated by NLW	4	10.0	7.5	10.0	10.0
	Percentage of encounters in which non-targeted infrastructure or materiel is incapacitated by NLW	2	7.5	7.5	10.0	10.0
	Timeline between NLW use and incapacitation	4	10.0	10.0	10.0	10.0

TABLE B.2
Output Metrics

Element	Metrics	Number of Vignettes Applicable (out of 13)	Average Validity	Average Reliability	Average Feasibility	Average Timeliness
Effectively responded to situations despite constraints	Percentage of tactical encounters in which use of NLWs was permissible, but lethal force was not	7	10.0	10.0	10.0	10.0
	Whether NLWs are allowed by ROE (binary yes/no distinction)	13	9.2	10.0	10.0	10.0
	Degree to which targeted populations perceive NLWs as equivalent to lethal weapons	13	9.6	5.0	5.8	4.6
Gained time before deciding to take lethal action	Time between initial use of NLWs and when a decision to authorize lethal force would have been required	5	10.0	9.0	10.0	10.0
	Percentage of encounters in which lethal action was not taken but would have been if NLWs were not available to delay decision	5	10.0	10.0	10.0	10.0

Table B.2—Continued

Element	Metrics	Number of Vignettes Applicable (out of 13)	Average Validity	Average Reliability	Average Feasibility	Average Timeliness
Gained time before deciding to take lethal action (continued)	Total interaction time between actors for interactions in which NLWs were used compared with those in which they were	6	10.0	10.0	10.0	10.0
	Commander's perception of increased decision time due to NLWs (yes/no)	6	10.0	10.0	10.0	10.0
	Time required to switch from nonlethal to lethal capability if EoF is necessary	5	10.0	10.0	10.0	10.0

Table B.2—Continued

Element	Metrics	Number of Vignettes Applicable (out of 13)	Average Validity	Average Reliability	Average Feasibility	Average Timeliness
Minimized collateral damage and fratricide	Percentage of tactical encounters in which there were *numerous* injuries among noncombatants	4	10.0	7.5	8.8	8.8
	Percentage of tactical encounters in which NLWs were used in which there were *any* serious/critical/(life/limb/sensory)/non-buddy care injuries among noncombatants relative to encounters in which NLWs were not used	4	10.0	8.8	8.8	8.8
	Percentage of tactical encounters in which NLWs were used in which there were fatalities among noncombatants relative to encounters in which NLWs were not used	4	10.0	8.8	10.0	10.0

Table B.2—Continued

Element	Metrics	Number of Vignettes Applicable (out of 13)	Average Validity	Average Reliability	Average Feasibility	Average Timeliness
Minimized collateral damage and fratricide (continued)	Average number of serious injuries among noncombatants per tactical encounter involving NLWs, relative to average number per tactical encounter not involving NLWs	4	10.0	7.5	6.3	7.5
	Average number of fatalities among noncombatants per tactical encounter involving NLWs, relative to average number per tactical encounter not involving NLWs	4	10.0	7.5	6.3	7.5
	Frequency and magnitude of long-term psychological effects of NLWs by targets	4	10.0	2.5	2.5	0.0
	Number of people unintentionally affected by NLW (accuracy/precision of NLW) per usage	3	10.0	5.0	6.7	8.3

Table B.2—Continued

Element	Metrics	Number of Vignettes Applicable (out of 13)	Average Validity	Average Reliability	Average Feasibility	Average Timeliness
Minimized collateral . . . (continued)	Frequency and severity of long-term biological effects of NLWs on targets	4	10.0	2.5	2.5	0.0
Conserved and augmented lethal capabilities	Percentage of tactical encounters in which lethal capabilities were not used	5	9.0	10.0	10.0	10.0
	Percentage of tactical encounters in which NLWs increased effectiveness of lethal weapons (e.g., enabled more selective targeting, less restrictive ROE)	4	10.0	8.8	10.0	10.0
Reduced risk of U.S., partner personnel casualties	Percentage of tactical encounters with U.S. and/ or partner casualties when NLWs were used relative to those when they were not	11	10.0	10.0	10.0	10.0
	Time required to switch from nonlethal to lethal capability if EoF is necessary	6	10.0	10.0	10.0	10.0

Table B.2—Continued

Element	Metrics	Number of Vignettes Applicable (out of 13)	Average Validity	Average Reliability	Average Feasibility	Average Timeliness
Reduced risk to U.S. systems or facilities	Percentage of tactical encounters with system casualties when NLWs were used relative to those when they were not	9	10.0	10.0	10.0	10.0
	Time required to switch from nonlethal to lethal capability if EoF is necessary	4	10.0	10.0	10.0	10.0
Reduced U.S. tactical costs (broadly defined)	Cost differential between use of NLWs and use of lethal systems (per use and fixed costs)	7	10.0	10.0	10.0	10.0
	Percentage reduction in capacity for lethal capabilities due to inclusion of NLWs in vehicles, vessels, backpacks, etc.	11	10.0	10.0	10.0	10.0
	Logistics (storage, transportation, resupply, etc.) requirements for NLWs relative to lethal systems	13	10.0	10.0	10.0	10.0

Table B.2—Continued

Element	Metrics	Number of Vignettes Applicable (out of 13)	Average Validity	Average Reliability	Average Feasibility	Average Timeliness
Reduced U.S. tactical costs . . . (continued)	Spare parts and maintenance requirements (time, cost, skill) of NLWs relative to lethal systems	13	10.0	10.0	10.0	10.0
Reduced risk of exceeding ROE or Laws of War	Percentage of encounters in which personnel mistakenly use NLWs in ways that accidentally exceed ROE or Laws of War	13	10.0	5.0	5.0	5.0
	Percentage of encounters in which personnel unnecessarily use NLWs in ways that exceed ROE or Laws of War	11	10.0	5.0	5.0	5.0
	Percentage of encounters in which personnel intentionally use NLWs in ways that exceed ROE or Laws of War	13	10.0	5.0	5.0	5.0

Table B.2—Continued

Element	Metrics	Number of Vignettes Applicable (out of 13)	Average Validity	Average Reliability	Average Feasibility	Average Timeliness
Reduced risk of exceeding ROE or Laws of War (continued)	Percentage of encounters in which use of NLWs is proportionate, whereas lethal force would have led to disproportionate/ indiscriminate effects	10	10.0	5.0	5.0	5.0
	Percentage of encounters in which NLWs enable compliance with ROE/ LOW, when lethal force would have resulted in exceeding/violations	10	10.0	5.0	5.0	5.0
Gathered intelligence from captured personnel and materiel	Percentage of encounters in which useful intelligence was gathered from personnel captured through use of NLWs vs. the same metric for lethal weapons	4	10.0	10.0	6.3	6.3
	Percentage of encounters in which useful intelligence was gathered from materiel captured through use of NLWs vs. the same metric for lethal weapons	5	10.0	10.0	10.0	10.0

Table B.2—Continued

Element	Metrics	Number of Vignettes Applicable (out of 13)	Average Validity	Average Reliability	Average Feasibility	Average Timeliness
Reduced adversary options and imposed costs	Change (absolute or percentage) in number of distinct options available to an adversary due to use of NLWs	13	10.0	5.4	10.0	10.0
	Percentage of encounters in which number of adversary options is reduced due to use of NLWs	13	10.0	5.4	10.0	10.0
	Percentage of encounters in which adversary experiences additional costs due to use of NLWs	13	10.0	5.4	10.0	10.0
Increased options for engaging targets	Change (absolute or percentage) in number of distinct options available due to use of NLWs	13	10.0	10.0	10.0	10.0
	Percentage of encounters in which number of options is increased due to use of NLWs	13	10.0	10.0	10.0	10.0

Table B.2—Continued

Element	Metrics	Number of Vignettes Applicable (out of 13)	Average Validity	Average Reliability	Average Feasibility	Average Timeliness
Enabled pre-emptive action without appearing to be an aggressor	Percentage of encounters in which pre-emptive action was taken using NLWs but would not have been with lethal systems due to risk of perception as aggressor	5	8.0	10.0	10.0	10.0
	Percentage of encounters in which pre-emptive action was not taken with either NLWs or lethal systems, because would have been perceived as aggressor with either	6	8.3	10.0	10.0	10.0
Enabled lower-signature clandestine operations	Signatures of NLW relative to alternative lethal system	0	0.0	0.0	0.0	0.0
	Attributability: Probability of being identified as U.S. operation with use of NLW, relative to without it	0	0.0	0.0	0.0	0.0

TABLE B.3
Outcome Metrics

Element	Metric	Number of Vignettes Applicable (out of 13)	Average Validity	Average Reliability	Average Feasibility	Average Timeliness
Competed effectively and demonstrated resolve while managing escalation in peacetime, gray-zone, and hybrid contexts	Percentage of incidents using NLWs that resulted in unwanted escalation divided by percentage of incidents not using NLWs that resulted in unwanted escalation	7	10.0	10.0	10.0	9.3
	Percentage of particular peacetime/ gray-zone/hybrid incidents in which NLWs were used	9	8.3	10.0	10.0	10.0
	Percentage of incidents in which NLWs were used and commanders perceived them as contributing effectively	9	10.0	10.0	10.0	10.0
	Degree to which targeted populations perceive NLWs as equivalent to lethal weapons	9	8.9	5.0	4.4	4.4

Table B.3—Continued

Element	Metric	Number of Vignettes Applicable (out of 13)	Average Validity	Average Reliability	Average Feasibility	Average Timeliness
Conducted operations in environments that were otherwise too dangerous due to collateral damage, fratricide, or escalation risks	Frequency of operations within a given time frame conducted with NLWs available that would not have been conducted without NLWs due to risks of collateral damage, fratricide, or escalation	3	10.0	10.0	10.0	10.0
Avoided alienation of population, host-nation forces, and host government	Host nation public opinion of use of NLWs, measured by polls	5	10.0	5.0	0.0	0.0
	Host nation public opinion of U.S. force presence and actions, measured by polls	5	10.0	5.0	0.0	0.0
	Frequency and scale of protests and riots against U.S. presence, actions	5	10.0	10.0	10.0	10.0

Table B.3—Continued

Element	Metric	Number of Vignettes Applicable (out of 13)	Average Validity	Average Reliability	Average Feasibility	Average Timeliness
Avoided alienation of population . . . (continued)	Frequency and scale of protests and riots related to events involving U.S. use of NLWs	5	10.0	6.0	10.0	10.0
	Frequency and scale of protests and riots against U.S. use of NLWs	5	10.0	5.0	10.0	10.0
	Degree of military cooperation/ permissiveness (high, medium, or low), as assessed by U.S. personnel	5	10.0	7.0	10.0	10.0
	Host nation forces' perception of U.S. use of NLWs, as assessed by U.S. personnel engaged with them	5	10.0	7.0	10.0	10.0
	Host government's perception of U.S. use of NLWs, as assessed by U.S. personnel engaged with them	5	10.0	7.0	10.0	10.0

Table B.3—Continued

Element	Metric	Number of Vignettes Applicable (out of 13)	Average Validity	Average Reliability	Average Feasibility	Average Timeliness
Avoided alienation of population . . . (continued)	Frequency of negative public statements by government figures about U.S. use of NLWs	5	10.0	8.0	10.0	10.0
	Degree to which targeted populations perceive NLWs as equivalent to lethal weapons	5	10.0	5.0	0.0	0.0
Enhanced perceptions of U.S. forces (in the United States and internationally)	Degree to which international public opinion perceives NLWs as equivalent to lethal weapons, measured by polls	12	10.0	5.0	0.0	0.0
	Degree to which U.S. public opinion perceives NLWs as equivalent to lethal weapons, measured by polls	12	10.0	5.0	0.0	0.0

Table B.3—Continued

Element	Metric	Number of Vignettes Applicable (out of 13)	Average Validity	Average Reliability	Average Feasibility	Average Timeliness
Enhanced perceptions of U.S. forces . . . (continued)	International public opinion of U.S. use of NLWs, measured by polls	12	10.0	5.0	0.0	0.0
	International public opinion of U.S. force presence and actions in a third country, measured by polls	7	10.0	5.0	0.0	0.0
	Frequency and scale of protests and riots internationally against U.S. presence, actions in third country	8	10.0	10.0	10.0	10.0
	U.S. public opinion of U.S. use of NLWs outside the United States, measured by polls	10	10.0	5.0	0.0	0.0
	U.S. public opinion of U.S. military use of NLWs domestically or along U.S. borders, measured by polls	2	10.0	5.0	0.0	0.0

Table B.3—Continued

Element	Metric	Number of Vignettes Applicable (out of 13)	Average Validity	Average Reliability	Average Feasibility	Average Timeliness
Enhanced perceptions of U.S. forces . . . (continued)	Frequency and scale of protests and riots in the United States against U.S. presence, actions in another country	8	10.0	10.0	10.0	10.0
	Frequency and scale of protests and riots related to events involving U.S. use of NLWs domestically or along U.S. borders	2	10.0	10.0	10.0	10.0
	Frequency and scale of U.S. protests and riots related to NLW use within the U.S. or along the border	2	10.0	5.0	10.0	10.0
	U.S. public opinion of U.S. force presence and actions in another country, measured by polls	6	10.0	5.0	0.0	0.0

Table B.3—Continued

Element	Metric	Number of Vignettes Applicable (out of 13)	Average Validity	Average Reliability	Average Feasibility	Average Timeliness
Increased partner cooperation	Degree of military cooperation/ permissiveness (high, medium, or low), as assessed by U.S. personnel	7	10.0	5.0	10.0	10.0
	Number of joint exercises, patrols, or other activities between U.S. and partner nation forces	7	10.0	10.0	10.0	10.0
Reused captured infrastructure and materiel	Timeline to repair after NLW usage relative to timeline imposed if needed to replace or use alternative	0	0.0	0.0	0.0	0.0
	Resource requirements to repair after NLW usage relative to resources required to replace or use alternative	0	0.0	0.0	0.0	0.0

Table B.3—Continued

Element	Metric	Number of Vignettes Applicable (out of 13)	Average Validity	Average Reliability	Average Feasibility	Average Timeliness
Avoided rebuilding costs	Timeline to repair after NLW usage relative to timeline imposed if needed to replace or use alternative	0	0.0	0.0	0.0	0.0
	Resource requirements to repair after NLW usage relative to resources required to replace or use alternative	0	0.0	0.0	0.0	0.0
Set standards for partner nations	Number of partner nations adopting NLWs and related tactics	12	10.0	5.4	5.4	5.0
	Number of partner nations violating Laws of War	9	10.0	5.0	5.0	5.0
	Number of partner nations found to have used NLWs for human-rights violations	8	10.0	10.0	10.0	10.0

Table B.3—Continued

Element	Metric	Number of Vignettes Applicable (out of 13)	Average Validity	Average Reliability	Average Feasibility	Average Timeliness
Reduced negative effects on morale from collateral damage or substantially harming individuals without lethal intent	Percentage of surveyed personnel who feel that NLWs reduce collateral damage	6	10.0	10.0	5.0	5.0
	Percentage of surveyed personnel who indicate that collateral damage contributed to negative morale	6	10.0	10.0	5.0	5.0
	Percentage of surveyed personnel who feel that they used lethal force in an unethical way	4	10.0	10.0	5.0	5.0
	Frequency of PTSD	6	0.0	0.0	0.0	0.0
	Frequency of suicide attempts	6	0.0	5.0	0.0	0.0

Notional Vignettes

To develop notional vignettes for the purpose of examining potential NLW usage, we had to ensure that we could confidently enumerate all relevant types of vignettes. We turned to previous RAND research on scenario design, which highlighted two considerations. First, NLWs may be useful across the spectrum of conflict, so there is a need to distinguish between structural and proximate factors of conflict when designing scenarios.[1] Second, non-military factors need to be strengthened and characterized in greater fidelity to improve the quality of military and political strategic analysis.[2]

Using these factors, we developed three design considerations that we varied to yield the relevant range of vignettes. The considerations, discussed in detail in this appendix, were whether

- the adversary sought to escalate the situation
- U.S. forces could feasibly withdraw
- the narrative surrounding the incident was stable.

Because these considerations are binary (e.g., a narrative is either stable or unstable), eight vignette combinations are possible. We built vignettes to include all of these combinations, using contemporary and past events as our guide. For example, consider the narrative with the combination of an escalatory adversary and U.S. forces that can withdraw and are unstable. We saw that incidents between U.S. and Russian ground forces in Syria fit that

---

[1] Heath and Lane, 2019, pp. 11–16.

[2] Heath and Lane, 2019, p. 6.

combination, so we built a vignette around it.[3] Each vignette reflects one or more contemporary, or recent, events, sometimes translated to another portion of the globe. These vignettes are intended solely for exploratory purposes; they are not meant to be predictive of future events.

The vignettes are assumed to take place in the mid-2020s, by which time all of the NLWs mentioned within them have been fully developed and are in use by operational forces, with appropriate CONOPS and tactics. From a geopolitical perspective, we assume that the world has not changed substantially since 2021. For the domestic vignettes (4 and 5), we also assume that there are still disgruntled individuals within the United States who are operating as part of self-described militias opposed to the government and many of their fellow citizens.

Table C.1 lists each vignette, together with its characteristics in terms of the adversary's penchant for escalation, the viability of withdrawal, and whether the narrative is stable.

In the remainder of this appendix, we provide a narrative describing each vignette, in the order listed in Table C.1. We begin with a general description of the vignette, followed by a characterization of both U.S. goals and those of the other party. We then present the potential results without NLWs, followed by an explanation of which NLWs could be used in the vignette and how they could be used.

Vignette 1: JSTARS Intercepted by PLANAF Fighters

General Description

The U.S. Army and Japanese Ground Self-Defense Force are conducting a Dynamic Force Employment exercise, in which they deliberately aim to be operationally unpredictable, on the Japanese island of Yonaguni in the East China Sea, supported by the U.S. Air Force and Japanese Air Self-Defense Force (see Figure C.1).

---

3 Schmitt, 2020.

TABLE C.1
Vignette Summary

Vignette	Domain	Services	GCC	Escalatory Adversary	Withdrawal Possible	Stable Narrative
1	JSTARS intercepted by PLANAF fighters	Air	USAF	INDOPACOM	X	X	X
2	Motorized confrontation with Russian forces in Syr a	Ground	Army	CENTCOM	X	X	
3	Boats approach a destroyer in the Strait of Gibraltar	Maritime	USN	EUCOM	X		X
4	Ground threat around domestic HA/DR sites	Ground	Army, USMC, National Guard	NORTHCOM	X		X
5	Maritime threat around domestic HA/DR sites	Maritime	USN, USCG	NORTHCOM	X		X
6	Securing embassy in Bahrain	Ground	USMC	CENTCOM	X		
7	Demonstrators in Palau	Ground	USAF	INDOPACOM	X		
8	Lasing in Palau	Ground	USAF	INDOPACOM	X		
9	Maritime standoff in the South China Sea	Maritime	USN, USCG	INDOPACOM		X	X
10	Maritime standoff in the Arctic	Maritime	USCG	EUCOM		X	X

Table C.1—Continued

Vignette	Domain	Services	GCC	Escalatory Adversary	Withdrawal Possible	Stable Narrative
11 Blockade enforcement near Venezuela	Maritime	USN	SOUTHCOM		X	X
12 SOF hostage rescue in Somalia	Ground	SOF	AFRICOM		X	
13 EAB defense against UAVs in the Philippines	Multi	USMC	INDOPACOM			

NOTES: A cell with an X means the vignette has that feature; an empty cell means the vignette does not have that feature.

AFRICOM = U.S. Africa Command; EUCOM = U.S. European Command; INDOPACOM = U.S. Indo-Pacific Command; NORTHCOM = U.S. Northern Command; SOUTHCOM = U.S. Southern Command; USAF = U.S. Air Force; USCG = U.S. Coast Guard; USMC = U.S. Marine Corps; USN = U.S. Navy.

FIGURE C.1
Location of Yonaguni

Due to a logistical delay, only one aircraft has arrived so far: a JSTARS aircraft, which conducts ground surveillance, battle management, command, and control. It is supporting the exercise north of Yonaguni when two PLANAF J-9II interceptor fighters begin to approach it. They make a series of close passes, while calling the JSTARS aircraft on standard radio channels, telling the pilot to get out of the area before they have to take drastic measures. They periodically turn fire-control radars on and off, and sometimes take turns operating just above the JSTARS, trying to force it to descend to avoid a collision.

Desired Goals

U.S. Goals

The United States wants to avoid a collision but also avoid having its aircraft pushed out of the area by the PLANAF aircraft.

PLANAF Aircraft Goals

The two PLANAF aircraft are trying to drive the U.S. aircraft out of the area, or perhaps to force it to land and remain on the ground.

Potential Results without NLWs

Repeated passes by the PLANAF aircraft could result in an accidental collision, which happened in April 2001 when a PLANAF aircraft collided with a U.S. Navy plane near Hainan Island, China. Alternatively, if the JSTARS backs away, that would be a demonstration of Chinese power even well offshore.

NLW Usage

The JSTARS pilot could use the LROI, a laser dazzler, to intermittently inflict glare to convey to the PLANAF pilots that they should back off. The pilot would want to be judicious about using it for fear that creating glare for too long could cause an accidental collision or crash.

Vignette 2: Motorized Confrontation with Russian Forces in Syria

General Description

U.S. and Russian forces are operating in northeast Syria in support of several armed factions vying for control. Russian forces do not view U.S. forces as enemy combatants, but territory controlled by each side's factions are commingled, U.S. and Russian convoys and patrols are operating close to each other, and tensions are rising.

Deconfliction is increasingly difficult as Russian patrols and convoys request permission to cross U.S.-controlled territory with greater frequency. Russian patrols have stopped waiting for permission to be given and are crossing U.S.-controlled territory, which has forced U.S. combat patrols to intercept Russian forces. During these encounters, Russian and U.S. forces (often motorized patrols) drive aggressively at each other, attempting to run vehicles off the road. On occasion, vehicles have collided, but no lethal weapons have been used. These incidents have occurred in areas where there are few civilians or other vehicles.

Russian patrols often consist of a mix of armored personnel carriers (APCs) (e.g., BTR-80s and Kamaz Typhoons), mine-resistant ambush protected (MRAPs), and occasionally rotary-wing assets (e.g., Mi-8 Hinds) that perform show-of-force passes at U.S. vehicles. U.S. patrols consist of MRAPs and MRAP all-terrain vehicles. All vehicles are armored and, with the exception of the APCs, are top-heavy and have relatively poor handling characteristics. All vehicles are armed with crew-served weapons (e.g., medium and heavy machine guns, automatic grenade launchers).

Desired Goals

U.S. Goals

The United States is attempting to conduct its operations in support of local forces without suffering casualties, provoking Russian forces, or forcing an escalatory dynamic that would require the United States to either commit more deeply to the Syrian conflict or withdraw.

Russian Goals

Russia believes itself to be ascendant in Syria and is looking to press the United States to cede territory or influence by continually testing, probing, and provoking U.S. responses to see what it can get away with. U.S. injuries, fatalities, or prominent withdrawals where U.S. forces lose face or credibility would be an ideal way to provoke a U.S. strategic response.

Potential Results Without NLWs

These road rage incidents are likely to continue, because Russian forces are keen on provoking a U.S. response. Most of the time, U.S. forces respond in measured ways, seeking to avoid collisions or temporarily withdrawing. However, these situations are fast-moving and unstable enough that injuries or fatalities might result from collisions, rollovers, or lethal fires.

NLW Usage

Acoustic hailers, including the AHD, could be used to warn Russian forces to back away and to avoid aggressive, reckless behaviors. Laser dazzlers, specifically an OI or an LROI, could be used to create an intense glare that would help further discourage aggressive behavior and would make it harder for drivers to approach at high speeds, or at all. Alternatively, the EoF CROWS system, which includes both hailing and dazzling capabilities, could be a substitute for the other systems. Finally, if Russian vehicles continued to try to approach, the PEVS or the RFVS could shut down their engines.

Vignette 3: Boats Approach a U.S. Destroyer in the Strait of Gibraltar

General Description

A U.S. destroyer is just west of the Strait of Gibraltar, headed for the Mediterranean. Three smaller vessels are nearby and appear to be on paths that converge with it, though this may just be indicative of the fact that they all are about to enter a narrow waterway. Two of them are speedboats, while one is larger, with an enclosed pilot house. People on these vessels appear to be shouting, shaking their fists, and making rude gestures at the destroyer. The destroyer continues in its path, then initiates use of an acoustic hailer to warn the vessels to back off, though they appear to be staying at roughly the same distance from the destroyer. Use of the LROI to dazzle the pilots seems to be making the vessels weave a bit more, but they persist in remaining relatively close. After a few moments, one of the speedboats appears to be gradually veering closer to the destroyer's path, but not quite making a beeline for it. The destroyer prepares to graduate to the use of force, perhaps by using warning munitions or the VIPER system to disable engines, before using lethal weapons.

Desired Goals

U.S. Goals

The United States wants to ensure that the boats do not threaten the destroyer (e.g., by colliding with it and detonating).

Smaller-Vessel Goals

The operators of the smaller vessels are trying to antagonize the destroyer, given their anti-American sentiments, but (unbeknownst to the destroyer) they are not trying to launch an actual attack. The destroyer also does not know that these are Moroccan nationals.

Potential Results Without NLWs

In the absence of NLWs, the destroyer will probably use lethal force. Killing unarmed Moroccans just outside Moroccan waters could lead to tensions with Morocco, as the shooting of an unarmed Egyptian by U.S. forces in the Suez Canal did in 2008.

NLW Usage

The destroyer's crew could initially use an acoustic hailer (e.g., AHD), LROI, or EoF CROWS to warn the boats away with a combination of hailing and dazzling. This could be followed by the VIPER system to disrupt their electronics or the MVSOT to stop their propellers, disabling them. Warning munitions might also be used. If these options failed to have the desired effect, the destroyer's crew would likely use lethal force.

Vignettes 4 and 5: Ground and Maritime Threats to Domestic HA/DR Efforts (Two Vignettes)

General Description

The most damaging hurricane in U.S. history has struck Louisiana and Mississippi, resulting in massive flooding, numerous people being trapped, shortages of supplies, and other calamities. DoD forces have been deployed to respond, alongside the Coast Guard, Federal Emergency Management Agency, and many other agencies. The Coast Guard and Navy are delivering supplies and helping with rescue along the Mississippi and in flooded areas. The Air Force is flying in supplies while also conducting medical evacuation for those who need it; both the Air Force and the U.S. Space Force are contributing to situational awareness. The Army, National Guard, Marine Corps, and reservists from multiple services are helping people who need shelter, food, and support, as are medics from all of the services.

However, members of a previously unknown, self-described militia—calling themselves *True Patriots*—are harassing relief efforts. They oppose all forms of federal authority, embrace an anarchistic ideology, and loathe minorities of various types. Although some True Patriots appear to be locally based, others have come from various states to engage in counter-government and counter–aid recipient activities. Some conspicuously armed members lurk around the periphery of temporary shelter and distribution facilities (both on foot and in vehicles), shouting their hatred of the government and their anger at people who would take help from it, threatening and intimidating aid recipients. Militia members are particularly focusing on aid facilities near where large populations of minorities live, and sometimes shout ethnic slurs and threats to "kill all of those people." Sometimes they "drill" right outside these facilities by having several individuals demonstratively load their weapons at once. Most of these activities are taking place just beyond the cordoned boundaries that are being secured by military personnel.

The True Patriots are also driving small open-top boats up to Navy and Coast Guard ships in the Mississippi and just offshore. At times, the personnel aboard these vessels have been not only shouting taunts and making rude gestures but also menacingly brandishing rifles and shotguns. Some have even fired into the air. There are no other vessels getting close to the Navy and Coast Guard ships.

Desired Goals

U.S. Government Goals

U.S. military personnel and civilian agencies aim to provide aid and mitigate the damage without disruption or intimidation of civilians, but also to avoid killing people unnecessarily.

Militia Goals

Members of the militia seem to be primarily interested in intimidating civilian populations, taunting military personnel, and potentially provoking violence, rather than using lethal force themselves. However, it is possible that members of this loosely organized group may actually use their weapons to try to kill civilians or military personnel.

Potential Results Without NLWs

In the absence of NLWs, there are three key risks. The first is that continued militia efforts could increasingly impede the ability of people who need aid to receive it and hinder overall response to the situation. The second is that militia members might actually shoot either civilians or military personnel, perhaps with minimal warning. The third is that military personnel from DoD services or the Coast Guard could use lethal force against militia members, which could lead to a series of violent attacks by the militia, the militia garnering traction due to its "martyrs," and the risk of collateral damage in areas where non–militia members are present.

NLW Usage

Both on land and along the rivers, a combination of acoustic hailers and laser dazzlers, perhaps using the integrated EoF CROWS or the separate AHD and LROI, could be used to discourage the militia members from approaching. An ADS could also be used selectively to deter militia members and get everyone to back off.

Navy and Coast Guard vessels dealing with the maritime threat could employ all of the above but would also have additional options: pepper balls and beanbag rounds or rubber bullets could also be used. (On land, these systems would not likely be chosen, because they could also affect populations seeking relief.) The VIPER could be used to electronically disable the boats, or the MVSOT could mechanically impede their ability to move. Although the Coast Guard routinely stops smuggling vessels by shooting out their engines, using VIPER or MVSOT would reduce the risk of escalating these incidents, because militia members might shoot back if fired upon (particularly if they did not realize that the engines were the target).

Vignette 6: Securing the Embassy in Bahrain

General Description

As the United States appears to be on the verge of war with Iran, the U.S. Embassy in Bahrain has been surrounded by huge crowds of people of all ages, demanding that the United States pull out of the Middle East. Children and old people are conspicuously present, and most of the crowd is peaceful. However, some men in their late teens and early 20s are throwing Molotov cocktails, and others are attempting to scale the embassy walls. (Aside: There was a real-world mob surrounding the embassy in 2002; a Bahraini died of injuries from a rubber bullet fired by the Bahraini security forces, contributing to widespread unrest and anti-Americanism in Bahrain thereafter.) The Bahraini security forces are trying to secure the embassy perimeter, disperse the crowd, and prevent entry into the building. However, they are being overwhelmed. The U.S. Marine Corps is primarily operating within the embassy but might need to help contribute to perimeter defense.

Desired Goals

U.S. Goals

The U.S. goal is to prevent entry into the embassy and harm to personnel or the building, and secondarily to disperse the crowd, but without causing death or permanent injury.

Adversary Goals

Most of the crowd wants to demonstrate peacefully, and perhaps to impede access to/from the building. Some members of the crowd, however, are trying to inflict harm against personnel or the building itself.

Potential Results Without NLWs

If the only option available is lethal force, the United States may either kill or permanently injure individuals, and/or receive damage due to Molotov cocktails or angry rioters entering the building.

NLW Usage

Acoustic hailers could help turn away the bulk of the crowd, and, if necessary, an ADS and pepper balls could also be used to disperse the crowd. If this did not suffice to dissuade the more-aggressive individuals throwing Molotov cocktails, then the use of beanbags, rubber bullets, and other NLW blunt impact munitions could help suppress the agitators and contribute to others backing away.

Vignette 7 and 8: Demonstrators and Lasing in Palau (Two Vignettes)

General Description

The United States is fighting a major war in Asia. The U.S. Air Force is employing a temporary base on the island of Babeldaob, the largest island in Palau (see Figure C.2), as part of an overall effort to disperse its basing throughout the region (making it less vulnerable to concentrated missile barrages against a handful of major bases).

The U.S. adversary in this fight is not only targeting the base in Babeldaob with missiles but also seeking to harass the base by subverting the local population and launching attacks from populated areas. This takes two different forms:

FIGURE C.2
Map of Palau

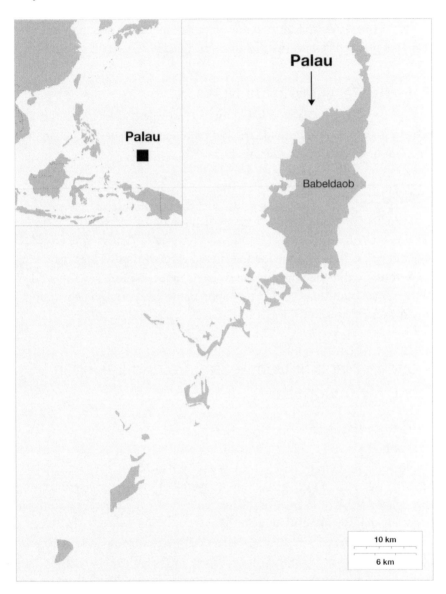

- Agents are paying families in the community to demonstrate in front of the base gates to block movements through them. Families providing pictures of their children participating in these demonstrations get extra money.
- The agents are also handing out high-powered laser pointers to local adolescents. Whenever they aim the pointers at the cockpits of aircraft that are landing or taking off (and prove it by having their friends take videos with their phones), they get money. This vignette is loosely based on reports of the use of high-powered laser pointers to target the cockpits of U.S. aircraft that are taking off or landing in Djibouti.

The base commander is struggling with how to deal with these threats. The crowds refuse to clear the base entrances, despite repeated imploring by heavily armed military police, and the local police force does not want to arrest families. The adolescents have been observed from the base's aircraft-control tower and by the UAVs the base uses for domain awareness, but they tend to scurry away before they can be apprehended, and lethal force is not authorized. Killing locals could lead to an anti-American backlash and cause the United States to lose access to the base.

Desired Goals

U.S. Goals

The United States is attempting to continue to operate aircraft, without meaningful disruption, from its base in Babeldaob. The demonstrations and laser attacks impede its ability to do so.

Adversary Goals

The adversary wants to disrupt air operations from the base, or halt them altogether, using several complementary approaches. Obstructing the base entrance prevents movement of personnel, equipment, and supplies; targeting pilots with laser pointers can damage their eyesight or compel the United States to limit (or even halt) operations. Moreover, if the United States can be provoked or tricked into harming local people, the adversary hopes that the Palau government will prevent it from continuing to operate from the base.

Potential Results Without NLWs

The use of lethal weapons against unarmed, peaceful families who are demonstrating, or against seemingly unarmed adolescents, would likely result in the United States losing access to the base. On the other hand, continuing impedance of access to the base would further disrupt operations, while damaging pilots' vision would impose both human and operational costs.

NLW Usage

Acoustic hailers could be used to encourage the demonstrators to disperse. Specific individuals in the crowd, especially those who appeared most aggressive, could be briefly targeted with an ADS to dampen their ardor and encourage them to leave. Finally, launching pepper balls would cause irritating sensations that would disperse the demonstrators.

Against the adolescents using powerful laser pointers, once their approximate location had been identified, roving patrols could use acoustic hailers or laser dazzlers (AHD, LROI, and/or EoF CROWS) to get them to cease their activity. The adolescents might try to aim their own, non-eye-safe laser pointers at the patrollers, so they should be wearing protective goggles, though it would be hard for the youth to accurately target the patrollers when they themselves were struggling to see. Beyond this, pepper balls or a vehicle-mounted ADS could be used (if the adolescents were particularly recalcitrant) to get them to cease their activities, make them easier to arrest, or perhaps simply to disperse them and deter further activity.

Vignette 9: Maritime Standoff in the South China Sea

General Description

A U.S. Navy destroyer is conducting freedom of navigation operations (FONOPs) near the Spratly Islands in the South China Sea; these islands, rocks, and reefs lie west of the Philippines and just north of the Malaysian portion of the island of Borneo (see Figure C.3). China Coast Guard (CCG) vessels approach, repeatedly swerving close to the destroyer and getting in its way, while contacting its bridge over the radio and telling the destroyer to

FIGURE C.3

Map of the South China Sea, Highlighting the Vicinity of the Spratly Islands

leave Chinese waters. (China claims full sovereignty over nearly the whole of the South China Sea, including the Spratly Islands; however, other nations dispute sovereignty over the islands, and under the United Nations Convention on the Law of the Sea, sailing close to the islands is permitted, regardless of which state has sovereignty over them.) The destroyer sends up a UAV to monitor the situation, though it does not closely approach the CCG ships. A People's Liberation Army Navy (PLAN) vessel approaches the group, at which point the CCG vessels become even more aggressive. The PLAN vessel's personnel ostentatiously begin to handle their machine guns and anti-aircraft guns and are periodically turning on the fire-control radar. They are aiming several handheld lasers at the destroyer's bridge and begin to take intermittent shots at the UAV.

Desired Goals

U.S. Goals

The destroyer does not want to be forced to leave the waterspace or to back down; it is also concerned about seeming to be insulted with impunity (e.g., by having its UAV shot down). At the same time, it does not want to escalate the situation, and it is also concerned about possible collisions or casualties.

Chinese Goals

China wants to push the U.S. Navy out of these waters and has a willingness to escalate if it thinks this is needed.

Potential Results Without NLWs

In the absence of NLWs, the United States could lose face if the destroyer is forced to leave the area, or if it absorbs insults without responding. The destroyer could also suffer equipment or even personnel casualties if it is unable to use any type of force. Alternatively, this situation could lead to dangerous escalation with China.

NLW Usage

LROI targeting the bridge could help warn CCG and PLAN ships to back off and could be followed up with an ADS to briefly target personnel on the deck to further underscore the message.

Vignette 10: Maritime Standoff in the Arctic

General Description

Russia has just announced that it has sovereign authority over most of the Arctic Ocean, expanding its previous claims. Previously, it had claimed total control of the Northern Sea Route between its Arctic islands and the mainland (contrary to the United Nations Convention on the Law of the Sea) plus extended continental shelf claims over the central Arctic Ocean (see Figure C.4). Now, it claims that nearly the entire Arctic Ocean constitutes its sovereign territory, drawing a "thirteen-dash line" over which it has the right to control all shipping and exclusive resource extraction rights, on the basis of historical Russian involvement in the region, without any basis in international law.

The United States, Canada, Denmark, and Norway want to counter Russia's claims through FONOPs. During the summer, they are demonstra-

FIGURE C.4
Map of the Arctic

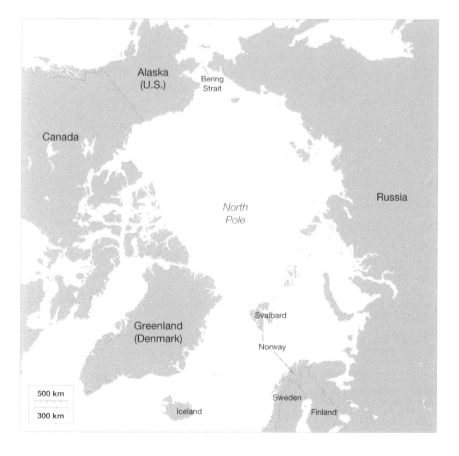

tively sending icebreakers through the central Arctic Ocean, where they will linger at the North Pole. The U.S. force consists of Coast Guard polar icebreakers because the U.S. Navy does not have those types of vessels. (Because this occurs in the mid-2020s, we assume that the U.S. Coast Guard has acquired some of the additional polar icebreakers that it currently plans to acquire; we also assume that they remain necessary, even in summer, because climate change has not yet eliminated year-round ice cover in the central Arctic Ocean.) Russia's nuclear-powered icebreakers are also confronting the NATO icebreakers near the North Pole, threatening to crash into them if they do not leave. The dynamics of the standoff at the North Pole are relatively slow, with icebreakers lumbering around as they crack through the frozen surface, and none of the icebreakers is heavily armed.

Desired Goals

U.S. and NATO Goals

The U.S. and NATO allies aim to prevent Russia from establishing, through precedent, that it has sovereignty over the Arctic Ocean. At the same time, they wish to avoid collisions or major escalation.

Russian Goals

Russia aims to drive NATO ships out of these waters without actual collisions or escalation. However, Russia is less risk-averse than NATO and may be willing to accept escalation if it happens.

Potential Results Without NLWs

In the absence of NLWs, a collision is a reasonable possibility. This could contribute to injuries, deaths, and escalation.

NLW Usage

The LROI could be used to dazzle the bridges of the Russian icebreakers, warning them to back off. In addition, brief engagements of an ADS could target personnel on the Russian icebreakers' decks to further underscore allied resolve and discourage the icebreaker from approaching more closely.

Vignette 11: Blockade Enforcement near Venezuela

General Description

The United States (including both Navy and Coast Guard vessels) and Brazil are leading a pan-American coalition blockading Venezuela until its military forces stop massacring civilians and take direction from the newly recognized government instead of the old regime. U.S. Navy, U.S. Coast Guard, Brazilian Navy, and other coalition forces are operating solely within international waters. However, Russian- and Chinese-flagged merchant ships, escorted by warships, are approaching Venezuela to flout the blockade, daring the United States and other nations to stop them. (This has similarities to the situation during the Cuban Missile Crisis, when the United States was trying to counter movements of Soviet vessels to Cuba.) The primary intent is to demonstrate that this is a so-called paper blockade, not a real one. The United States is seeking to counter these maritime flows without causing either civilian casualties (which could derail the coalition) or escalation with rival powers.

Desired Goals

U.S. Goals

The United States aims to prevent the Russian and Chinese merchant ships, including the warships escorting them, from going into or out of Venezuela but without causing major great-power escalation. It also aims to prevent smaller civilian vessels from smuggling goods but without causing permanent injuries or fatalities.

Adversary Goals

Russia and China want to show that the blockade is ineffective, and that the United States lacks resolve; they also want to help the government of Venezuela survive.

Potential Results Without NLWs

Using lethal weapons against Russian or Chinese ships could lead to dangerous escalation, while not acting could embolden those nations and reduce confidence in the coalition.

NLW Usage

LROIs could be used to dazzle the bridges of Russian and Chinese vessels, reminding them that the United States and other members of the coalition are monitoring their movements, adding resolve to the tone of direct bridge-to-bridge communications.

Vignette 12: SOF Hostage Rescue in Somalia

General Description

A group of 12 merchant sailors from a U.S.-flagged vessel have been kidnapped by Somali pirates for ransom in the Gulf of Aden. The pirates, knowing that the United States and other nations have successfully rescued kidnapped sailors being held in coastal areas, have moved the sailors to Haradhere, approximately 20 miles inland from the coastal town of Faax.

Surveillance indicates that sailors are being held in a compound; pattern of life observations suggest that 20 to 30 people are often in the building at a time; some are pirates, and others are regional tribal members with unclear affiliations who may be noncombatants.

The United States has received reports that several of the hostages are in declining health. Negotiators are working with the pirates' tribes to end the situation peacefully, but little progress is being made. The Somali government is receptive to a raid to rescue the hostages, but inadvertent deaths of the hostages or tribal members will adversely affect future hostage negotiations with pirates who continue to operate in the region. Joint Special Operations Command planners believe that a heliborne SOF raid conducted with intermediate force options can succeed while addressing Somali government concerns.

Desired Goals

U.S. Goals

The main U.S. goal is to rescue the merchant sailors. Additional considerations include minimizing collateral damage, avoiding civilian casualties, and reducing any negative effects of a violent, high-profile raid on the legitimacy of the Somali government and U.S. reputation.

Pirate Goals

The pirates are unlikely to want to kill the merchant sailors. Their goal is to receive a ransom payment.

Potential Results Without NLWs

Without NLWs, the Somali government may balk at the prospect of allowing U.S. troops to conduct operations in Mogadishu. Even if the government does allow it, the reputational risks and the potential for civilian casualties and collateral damage are considerable.

NLW Usage

The SOF could use a combination of NLWs to temporarily incapacitate the pirates while reducing the risks to the hostages, to themselves, and to noncombatants in the environment. Flash-bang grenades would generate intense light and sound to distract and briefly incapacitate individuals. Launching beanbag rounds, rubber bullets, and grenades that disperse rubber pellets could temporarily incapacitate individuals while minimizing the risk of civilian casualties and limiting collateral damage compared with the use of standard ammunition.

Vignette 13: EAB Defense Against UAVs in the Philippines

General Description

Tensions have increased between the Philippines and China over repeated clashes between Philippine merchant vessels and Chinese Coast Guard and naval militia vessels over disputed territories. The United States is supporting the Philippines by increasing is operational tempo in the region to signal to the Chinese that their actions against a U.S. treaty ally will not go unchallenged.

A Marine Littoral Regiment (MLR) operating in the area has been tasked with preventing the PLAN from further escalating the situation. To accomplish this, the MLR has very publicly deployed several EABs to battle positions at key choke points, including one equipped with naval strike missiles (NSMs). This EAB consists of two joint light tactical vehicles armed with NSMs, associated fire control elements, and a reinforced infantry platoon for security.

The PLAN is not willing to be targeted by this EAB but is not willing to destroy the EAB with military assets. Therefore, naval militia vessels have started to approach the EAB's location, knowing that the United States would not be willing to escalate by firing on an ostensibly civilian vessel. The naval militia vessels are launching small UAVs, some of which have electronic warfare capabilities that could neutralize the NSM fire control systems at close range.

Desired Goals

U.S. Goal

Continue to prevent the PLAN from entering the disputed area and escalating the situation by holding those ships under threat by NSMs launched from the EAB.

Adversary Goal

Neutralize the EAB to allow the PLAN to interject itself into the disputed area.

Potential Results Without NLWs

Without NLWs that can neutralize the naval militia ships and their small UAV, the EAB is vulnerable to being neutralized. Although ROEs allow it, the United States is not willing to cede the information battle to the Chinese by employing high-profile, lethal weapons against civilian merchant vessels.

NLW Usage

In this situation, targeting the UAVs with bursts of microwave energy—such as a version of the VIPER or a modified version of the RFVS that could be aimed at UAVs—would disable their electronics and cause them to crash into the sea.

Interview Protocols

We used three semistructured interview protocols: one for program officials, another for technologists, and one for end users; all are listed below.

Program Protocol

Background Information

1. What is your current rank, component, billet and military operational specialty, or (if civilian/contractor) job code?
2. What organization do you belong to?

Requirements

3. What experience have you had with NLWs?
4. Looking back, have there been situations where you wish you had NLWs?
5. How has DoD determined its needs for NLWs?
6. What metrics has DoD used to determine its need for NLWs?
7. What attributes make NLWs attractive or less attractive to you?
8. What challenges have you had with NLWs?
9. What organizational impediments have you had with NLWs?
10. What policy or doctrinal issues have you had with NLWs?

Future Use

11. Looking ahead, in what missions and operating environments do you expect NLWs to be used in the future?
12. What characteristics would make NLWs a suitable tool in these situations?
13. What implications do future operating concepts have on the demand for and utility of NLWs?
14. Future operating concepts include Multi-Domain Operations (MDO), Expeditionary Advanced Base Operations (EABO), agile combat employment (ACE), Distributed Mission Operations (DMO), etc.
15. What changes need to be made to enable NLWs' roles in future operating concepts?
16. What technical or tactical changes?
17. What organizational impediments must be overcome?
18. What policy or doctrinal issues must be addressed?
19. What are the most promising emerging NLW technologies?
20. Why?

Conclusion

21. Do you have anything to add that we did not cover in today's interview?
22. Do you have any recommendation of people we should speak with?

Technologist Protocol

Background Information

1. What is your current rank, component, billet and military operational specialty, or (if civilian/contractor) job code?
2. What organization do you belong to?
3. Requirements
4. What experience have you had with NLWs?

5. Looking back, have there been situations where you wish you had NLWs?
6. How has DoD determined its needs for NLWs?
7. What metrics has DoD used to determine its need for NLWs?
8. What attributes make NLWs attractive or less attractive to you?
9. What challenges have you had with NLWs?
10. What organizational impediments have you had with NLWs?
11. What policy or doctrinal issues have you had with NLWs?

Future Use

12. Looking ahead, in what missions and operating environments do you expect NLWs to be used in the future?
13. What characteristics would make NLWs a suitable tool in these situations?
14. What implications do future operating concepts have on the demand for and utility of NLWs?
15. Future operating concepts include MDO, EABO, ACE, DMO, etc.
16. What changes need to be made to enable NLWs' roles in future operating concepts?
17. What technical or tactical changes?
18. What organizational impediments must be overcome?
19. What policy or doctrinal issues must be addressed?
20. What are the most promising emerging NLW technologies?
21. Why?

Conclusion

22. Do you have anything to add that we did not cover in today's interview?
23. Do you have any recommendation of people we should speak with?

End User Protocol

Background Information

1. What is your current rank, component, billet and military operational specialty, or (if civilian/contractor) job code?
2. What organization do you belong to?

Past Use

3. What experience have you had with NLWs?
4. Looking back, have there been situations where you wish you had NLWs?
5. What attributes make NLWs attractive or less attractive to you?
6. Looking back, have there been situations where NLWs failed to fulfill their potential?
7. What challenges have you had with NLWs?
8. What technical or tactical challenges have you had with NLWs?
9. What organizational impediments have you had with NLWs?
10. What policy or doctrinal issues have you had with NLWs?

Future Use

11. Looking ahead, in what missions and operating environments do you expect NLWs to be used in the future?
12. What characteristics would make NLWs a suitable tool in these situations?
13. What implications do future operating concepts have on the demand for and utility of NLWs?
14. Future operating concepts include MDO, EABO, ACE, DMO, etc.
15. What changes need to be made to enable NLWs' roles in future operating concepts?
16. What technical or tactical changes?
17. What organizational impediments must be overcome?
18. What policy or doctrinal issues must be addressed?

Conclusion

19. Do you have anything to add that we did not cover in today's interview?

20. Do you have any recommendation of people we should speak with?

Abbreviations

ACE	agile combat employment
ADS	Active Denial System
AHD	acoustic hailing device
APC	armored personnel carrier
CCG	China Coast Guard
CONOPS	concept of operations
CROWS	Common Remotely Operated Weapons Station
C-UAV	counter-unmanned aerial vehicle
DMO	Distributed Mission Operations
DoD	U.S. Department of Defense
EAB	Expeditionary Advanced Base
EABO	Expeditionary Advanced Base Operations
EoF	Escalation of Force
FONOP	freedom of navigation operation
HA/DR	Humanitarian Assistance and Disaster Response
IFC	Intermediate Force Capability
JIFCO	Joint Intermediate Force Capabilities Office
JLTV	Joint Light Tactical Vehicle
JSTARS	Joint Surveillance and Target Attack Radar System
LROI	Long-Range Ocular Interrupter
MDO	Multi-Domain Operations
MLR	Marine Littoral Regiment
MRAP	mine-resistant ambush protected
MVSOT	Maritime Vessel Stopping Occlusion Technologies

N/A	not applicable
NATO	North Atlantic Treaty Organization
NLW	non-lethal weapon
NSM	Naval Strike Missile
OI	Ocular Interrupter
OSD	Office of the Secretary of Defense
PEVS	Pre-Emplaced Electric Vehicle Stopper
PLAN	People's Liberation Army Navy
PLANAF	People's Liberation Army Navy and Air Force
RFVS	Radio Frequency Vehicle Stopper
ROE	rules of engagement
SNS-RDD	Single Net Solution-Remote Deployment Device
SOF	special operations forces
TTPs	tactics, techniques, and procedures
UAS	unmanned aircraft system
UAV	unmanned aerial vehicle
VIPER	Vessel Incapacitating Power Effect Radiation

Bibliography

Air Force Instruction 31-118, *Security: Security Forces Standards and Procedures*, Washington, D.C., August 18, 2020.

Allison, Graham T., Paul X. Kelley, and Richard L. Garwin, *Nonlethal Weapons and Capabilities: Report of an Independent Task Force Sponsored by the Council on Foreign Relations*, New York: Council on Foreign Relations, February 2004. As of October 4, 2021:
https://cdn.cfr.org/sites/default/files/report_pdf/Nonlethal_TF.pdf

Alpert, Geoffrey P., Michael R. Smith, Robert J. Kaminski, Lorie A. Fridell, John MacDonald, and Bruce Kubu, *Police Use of Force, Tasers and Other Less-Lethal Weapons*, Washington D.C., National Institute of Justice, May 2011. As of October 4, 2021:
https://www.ojp.gov/pdffiles1/nij/232215.pdf

Army Nonlethal Scalable Effects Center, "Army Nonlethal Weapons APBI," presentation, September 21, 2017.

Bagwell, Randall, "The Treatment Assessment Process (TAP): The Evolution of Escalation of Force," *Army Lawyer*, No. DA PAM 27-50-419, 2008, pp. 5–17.

Beck, Jamal, "New Vehicle Stopper Trials Underway at Tinker Air Force Base," press release, Joint Non-Lethal Weapons Directorate, August 15, 2018.

Bedard, E. R., "Nonlethal Capabilities: Realizing the Opportunities," *Defense Horizons*, March 2002, pp. 1–6.

Berger, David H., "The Case for Change: Meeting the Principal Challenges Facing the Corps," *Marine Corps Gazette*, June 2020.

Berger, Miriam, "'Nonlethal Weapons' Fired at Protesters in Hong Kong, Chile and Iraq Are Having Very Dangerous—Even Deadly—Effects," *Washington Post*, November 23, 2019.

Bilms, Kevin, "Will COVID Finally Force Us to Think Differently About National Security?" *Defense One*, December 15, 2020. As of October 4, 2021:
https://www.defenseone.com/ideas/2020/12/will-covid-finally-force-us-think-differently-about-national-security/170781/

Burgei, Wesley A., Shannon E. Foley, and Scott M. McKim, "Developing Non-Lethal Weapons, The Human Effects Characterization Process," *Defense AT&L*, May–June 2015. As of October 6, 2021:
https://www.dau.edu/library/defense-atl/DATLFiles/May-Jun2015/Burgei%20et%20all.pdf

Chairman of the Joint Chiefs of Staff Instruction 3121.01B, *Standing Rules of Engagement (SROE)/Standing Rules for the Use of Force (SRUF) for U.S. Forces*, Washington, D.C., June 13, 2005.

Choudhury, Fareed, "Nonlethal Munitions (NLM) Expand Warfighter Capabilities," *Army AT&L*, January–March 2008, pp. 47–49. As of October 6, 2021:
https://asc.army.mil/docs/pubs/alt/2008/1_JanFebMar/articles/46_Nonlethal_Munitions_(NLM)_Expand_Warfighter_Capabilities_200801.pdf

Ciullo, Dan, Jeff deLongpre, Sim Mcarthur, Jake Nowakowski, Rich Shene, Earvin Taylor, Roosevelt White, Po-Yu Cheng, Yinghui Heng, Chia Sern Wong, Wai Keat Wong, Yee Ling Phua, Philip Zlatsin, Junwei Choon, Yong Shern Neo, Daryl Lee, Wen Chong Chow, Guan Hock Lee, Valentine Leo, Zhifeng Lim, Boon Chew Sheo, Sze Shiang Soh, Harn Chin Teo, and Sea Engineering Analysis Cohort SEA-19B, *Viable Short-Term Directed Energy Weapon Naval Solutions: A Systems Analysis of Current Prototypes*, Monterey, Calif.: Naval Postgraduate School, June 2013. As of October 5, 2021:
http://hdl.handle.net/10945/34734

CJCSI—*See* Chairman of the Joint Chiefs of Staff Instruction.

Clark, Bryan, Mark Gunzinger, and Jesse Sloman, *Winning the Gray Zone: Using Electromagnetic Warfare to Regain Escalation Dominance*, Washington, D.C.: Center for Strategic and Budgetary Assessments, CSBA6305, 2017.

Clark, Bryan, "China Already Has Escalation Dominance," *Inkstone*, January 4, 2019. As of October 5, 2021:
https://www.inkstonenews.com/opinion/bryan-clark-china-has-already-established-escalation-dominance-south-china-sea/article/2180184

Cleveland, Charles T., and Daniel Egel, *The American Way of Irregular War: An Analytical Memoir*, Santa Monica, Calif.: RAND Corporation, PE-A301-1, 2020. As of October 5, 2021:
https://www.rand.org/pubs/perspectives/PEA301-1.html

Connable, Ben, *Embracing the Fog of War: Assessment and Metrics in Counterinsurgency*, Santa Monica, Calif.: RAND Corporation, MG-1086-DOD, 2012. As of as of July 31, 2020:
https://www.rand.org/pubs/monographs/MG1086.html

Connable, Ben, Stephanie Young, Stephanie Pezard, Andrew Radin, Raphael S. Cohen, Katya Migacheva, and James Sladden, *Russia's Hostile Measures: Combating Russian Gray Zone Aggression Against NATO in the Contact, Blunt, and Surge Layers of Competition*, Santa Monica, Calif.: RAND Corporation, RR-2539-A, 2020. As of October 5, 2021:
https://www.rand.org/pubs/research_reports/RR2539.html

Coppernoll, Margaret-Anne, "The Nonlethal Weapons Debate," *Naval War College Review*, Vol. 52, No. 2, pp. 112–131, Spring 1999.

D'Andrea, John, Donald Cox, P. Henry, J. Ziriax, D. Hatcher, and W. Hurt, *Rhesus Monkey Aversion to 94-GHz Facial Exposure*, Brooks City Base, Tex.: Naval Health Research Center Detachment, Directed Energy Bioeffects Laboratory, Technical Report–NHRC DEBL TR-2006-07, September 2008.

Dedoose, version 8.0.35, web application for managing, analyzing, and presenting qualitative and mixed method research data, Los Angeles: SocioCultural Research Consultants, 2018.

Defense Science Board, *2019 DSB Summer Study on the Future of U.S. Military Superiority*, Washington, D.C.: Office of the Secretary of Defense, June 2020.

Department of Defense Directive 3000.03E, *DoD Executive Agent for Non-Lethal Weapons (NLW), and NLW Policy*, Washington, D.C., April 25, 2013, Incorporating Change 2, August 31, 2018.

Department of Defense Instruction 3200.19, *Non-Lethal Weapons (NLW) Human Effects Characterization*, Washington, D.C., May 17, 2012, Incorporating Change 1, September 13, 2017.

Department of Defense Instruction 3200.19, *Non-Lethal Weapons (NLW) Human Effects Characterization*, Washington, D.C., May 17, 2012, Incorporating Change 2, August 31, 2018.

Dobbins, James, Stephen Watts, Nathan Chandler, Derek Eaton, and Stephanie Pezard, *Seizing the Golden Hour: Tasks, Organization, and Capabilities Required for the Earliest Phase of Stability Operations*, Santa Monica, Calif.: RAND Corporation, RR-2633-A, 2020. As of September 28, 2021: https://www.rand.org/pubs/research_reports/RR2633.html

Duncan, James C., "A Primer on the Employment of Non-Lethal Weapons," *Naval Law Review*, Vol. 45, 1998, pp. 1–56.

Eckstein, Megan, "Pentagon's Non-Lethal Weapons Office Pushing Gray-Zone Warfare Tools," *USNI News*, September 24, 2019. As of October 5, 2021: https://news.usni.org/2019/09/24/pentagons-non-lethal-weapons-office-pushing-gray-zone-warfare-tools

Elder, R. Wyn, *The Role of Non-Lethal Air Power in Future Peace Operations: "Beyond Bombs on Target,"* Maxwell Air Force Base, Ala.: Air University Press, 2003.

Ellis, J. D., *Directed-Energy Weapons: Promise and Prospects*, Washington, D.C.: Center for a New American Security, 2015. As of October 5, 2021: https://books.google.com/books?id=nbyKrgEACAAJ

Flanagan, Stephen, Jan Osburg, Anika Binnendijk, Marta Kepe, and Andrew Radin, *Deterring Russian Aggression in the Baltic States Through Resilience and Resistance*, Santa Monica Calif.: RAND Corporation, RR-2779-OSD, 2019. As of October 5, 2021:
https://www.rand.org/pubs/research_reports/RR2779.html

Ford, Gerald, "Renunciation of Certain Uses in War of Chemical Herbicides and Riot Control Agents," Washington, D.C.: Executive Office of the President, Executive Order 11850, April 8, 1975. As of October 28, 2021:
https://www.archives.gov/federal-register/codification/executive-order/11850.html

Fowers, Alyssa, Aaron Steckelberg, and Bonnie Berkowitz, "A Guide to the Less-Lethal Weapons that Law Enforcement Uses Against Protesters," *Washington Post*, June 5, 2020. As of April 12, 2021:
https://www.washingtonpost.com/nation/2020/06/05/less-lethal-weapons-protests/?arc404=true

Freier, Nathan, and Jonathan Dagle, "The Weaponization of Everything," *Defense One*, September 9, 2018. As of October 5, 2021:
https://www.defenseone.com/ideas/2018/09/weaponization-everything/151097/

Fridman, Ofer, "A Technological Gap or Misdefined Requirements?" *Joint Force Quarterly*, Vol. 76, First Quarter 2015. As of October 5, 2021:
https://ndupress.ndu.edu/Publications/Article/577592/nonlethal-weapons-a-technological-gap-or-misdefined-requirements/

Frost, Gerald P., and Calvin Shipbaugh, *GPS Targeting Methods for Non-Lethal Systems*, Santa Monica, Calif.: RAND Corporation, RP-262, 1996. As of October 5, 2021:
https://www.rand.org/pubs/reprints/RP262.html

Gain, Nathan, "US Navy Lab Investigates Innovative Non-Lethal Boat Stopping Technology," *Naval News*, November 25, 2019. As of April 30, 2021:
https://www.navalnews.com/naval-news/2019/11/us-navy-lab-investigates-innovative-non-lethal-boat-stopping-technology/.

Glenn, Russell W., Sidney W. Atkinson, Michael Barbero, Frederick J. Gellert, Scott Gerwehr, Steven Hartman, Jamison Jo Medby, Andrew O'Donnell, David Owen, and Suzanne Pieklik, *Ready for Armageddon: Proceedings of the 2001 RAND Arroyo-U.S. Army ACTD-CETO-USMC Non-Lethal and Urban Operations Program Urban Operations Conference*, Santa Monica, Calif.: RAND Corporation, CF-179-A, 2002. As of September 28, 2021:
https://www.rand.org/pubs/conf_proceedings/CF179.html

Gompert, David C., John Gordon IV, Adam R. Grissom, David R. Frelinger, Seth G. Jones, Martin C. Libicki, Edward O'Connell, Brooke Stearns Lawson, and Robert E. Hunter, *War by Other Means: Building Complete and Balanced Capabilities for Counterinsurgency: RAND Counterinsurgency Study—Final Report*, Santa Monica, Calif.: RAND Corporation, MG-595/2-OSD, 2008. As of October 05, 2021:
https://www.rand.org/pubs/monographs/MG595z2.html

Gompert, David C., Stuart Johnson, Martin C. Libicki, David R. Frelinger, John Gordon IV, Raymond Smith, and Camille A. Sawak, *Underkill: Scalable Capabilities for Military Operations amid Populations*, Santa Monica, Calif.: RAND Corporation, MG-848-OSD, 2009. As of October 05, 2021:
https://www.rand.org/pubs/monographs/MG848.html

Granger, Dewey A., *Integration of Lethal and Nonlethal Fires: The Future of the Joint Fires Cell*, Fort Leavenworth, Ky.: School of Advanced Military Studies, U.S. Army Command and General Staff College, 2009. As of October 5, 2021:
https://apps.dtic.mil/sti/pdfs/ADA505041.pdf

Guest, Greg, Kathleen M. MacQueen, and Emily E. Namey, *Applied Thematic Analysis*, Thousand Oaks, Calif.: Sage Publications, 2011.

Gunzinger, Mark, and Christopher Dougherty, *Changing the Game: The Promise of Directed-Energy Weapons*, Washington, D.C.: Center for Strategic and Budgetary Assessments, 2012.

Hayes, Bradd C., *Naval Rules of Engagement: Management Tools for Crisis*, Santa Monica, Calif.: RAND Corporation, N-2963-CC, 1989. As of October 5, 2021:
https://www.rand.org/pubs/notes/N2963.html

Headquarters, Department of the Army, and Headquarters, U.S. Marines, *The Commander's Handbook on the Law of Land Warfare*, Quantico, Va.: Field Manual 6-27/Marine Corps Tactical Publication 11-10C, C1, August 2019.

Heath, Timothy R., and Matthew Lane, *Science-Based Scenario Design: A Proposed Method to Support Political-Strategic Analysis*, Santa Monica, Calif.: RAND Corporation, RR-2833-OSD, 2019. As of July 27, 2020:
https://www.rand.org/pubs/research_reports/RR2833.html

Higgins, Brian, "Final Findings from the Expert Panel on the Safety of Conducted Energy Devices," *NIJ Journal*, No. 268, 2011, pp. 32–35. As of October 5, 2021:
https://www.ojp.gov/pdffiles1/nij/235894.pdf

Hughes, Edward, ed., *Less-Lethal Operational Scenarios for Law Enforcement*, University Park, Pa.: Institute for Non-Lethal Defense Technologies, Applied Research Laboratory, The Pennsylvania State University, August 2005. As of October 5, 2021:
https://nij.ojp.gov/library/publications/
less-lethal-operational-scenarios-law-enforcement

Ioffe, Julia, "The Mystery of the Immaculate Concussion," *GQ*, October 19, 2020. As of October 6, 2021:
https://www.gq.com/story/cia-investigation-and-russian-microwave-attacks

Jerothe, Douglas J., *DoD Non-Lethal Weapons Program: Overview Brief and Information Exchange*, Joint Intermediate Force Capabilities Office, U.S. Department of Defense, Non-Lethal Weapons Program presentation to Keystone Course, National Defense University, Quantico, Va., 2015. As of October 5, 2021:
https://jnlwp.defense.gov/Portals/50/Documents/Resources/Presentations/
Overview_Presentations/Keystone%20_Brief_15Jan2015_logo_fix.pdf

JIFCO, DoD, Non-Lethal Weapons Program—*See* Joint Intermediate Force Capabilities Office, U.S. Department of Defense, Non-Lethal Weapons Program.

Joint Intermediate Force Capabilities Office, U.S. Department of Defense, Non-Lethal Weapons Program, "Active Denial System FAQs," webpage, undated-a. As of April 22, 2021:
https://jnlwp.defense.gov/About/Frequently-Asked-Questions/
Active-Denial-System-FAQs/

Joint Intermediate Force Capabilities Office, U.S. Department of Defense, Non-Lethal Weapons Program, "Human Electro-Muscular Incapacitation FAQs," webpage, undated-b. As of April 30, 2021:
https://jnlwp.defense.gov/About/Frequently-Asked-Questions/
Human-Electro-Muscular-Incapacitation-FAQs/

Joint Intermediate Force Capabilities Office, U.S. Department of Defense, Non-Lethal Weapons Program, "Oleoresin Capsicum Dispensers," webpage, undated-c. As of April 30, 2021:
https://jnlwp.defense.gov/Current-Intermediate-Force-Capabilities/
Oleoresin-Capsicum-Dispensers/

Joint Intermediate Force Capabilities Office, U.S. Department of Defense, Non-Lethal Weapons Program, "Variable Kinetic System (VKS) Non-Lethal Launcher and U.S. Coast Guard Pepperball Launcher Systems," webpage, undated-d. As of April 30, 2021:
https://jnlwp.defense.gov/Current-Intermediate-Force-Capabilities/
Variable-Kinetic-System/

Joint Intermediate Force Capabilities Office, U.S. Department of Defense, Non-Lethal Weapons Program, *Escalation of Force Options Annual Review*, Quantico, Va.: Non-Lethal Weapons Program, 2009a.

Joint Intermediate Force Capabilities Office, U.S. Department of Defense, Non-Lethal Weapons Program, "Initial Capabilities Document for Counter-Materiel Joint Non-Lethal Effects," 2009b.

Joint Intermediate Force Capabilities Office, U.S. Department of Defense, Non-Lethal Weapons Program, "Initial Capabilities Document for Counter-Personnel Joint Non-Lethal Effects," 2009c.

Joint Intermediate Force Capabilities Office, U.S. Department of Defense, Non-Lethal Weapons Program, *Non-Lethal Weapons for Today's Operations Annual Review*, Quantico, Va., 2010–2011.

Joint Intermediate Force Capabilities Office, U.S. Department of Defense, Non-Lethal Weapons Program, *Non-Lethal Capabilities for Complex Environments Annual Review*, Quantico, Va., 2012.

Joint Intermediate Force Capabilities Office, U.S. Department of Defense, Non-Lethal Weapons Program, *Non-Lethal Capabilities for Complex Environments Annual Review*, Quantico, Va., 2013.

Joint Intermediate Force Capabilities Office, U.S. Department of Defense, Non-Lethal Weapons Program, *DoD Non-Lethal Capabilities: Enhancing Readiness for Crisis Response Annual Review*, Quantico, Va., 2015. As of October 5, 2021:
https://jnlwp.defense.gov/Press-Room/Annual-Reviews/

Joint Intermediate Force Capabilities Office, U.S. Department of Defense, Non-Lethal Weapons Program, "Non-Lethal Optical Distracters Fact Sheet," May 2016a.

Joint Intermediate Force Capabilities Office, U.S. Department of Defense, Non-Lethal Weapons Program, *Strategic Plan 2016-2025, Science & Technology, Joint Non-Lethal Weapons Program*, Quantico, Va., 2016b.

Joint Intermediate Force Capabilities Office, U.S. Department of Defense, Non-Lethal Weapons Program, "Acoustic Hailing Devices Fact Sheet," November 16, 2018a.

Joint Intermediate Force Capabilities Office, U.S. Department of Defense, Non-Lethal Weapons Program, "Radio Frequency Vehicle Stopper," November 16, 2018b.

Joint Intermediate Force Capabilities Office, U.S. Department of Defense, Non-Lethal Weapons Program, "Single Net Solution with Remote Deployment Device," November 16, 2018c.

Joint Intermediate Force Capabilities Office, U.S. Department of Defense, Non-Lethal Weapons Program, "Vessel-Stopping Prototype," November 16, 2018d.

Joint Intermediate Force Capabilities Office, U.S. Department of Defense, Non-Lethal Weapons Program, *Intermediate Force Capabilities: Bridging the Gap Between Presence and Lethality, Executive Agent's Planning Guidance 2020*, March 2020a. As of April 20, 2021:
https://mca-marines.org/wp-content/uploads/DoD-NLW-EA-Planning-Guidance-March-2020.pdf

Joint Intermediate Force Capabilities Office, U.S. Department of Defense, Non-Lethal Weapons Program, Joint Integrated Product Team Chairman, Joint Integrated Product Team (JIPT) 20-01 Summary of Conclusions, Quantico, Va., March 4, 2020b.

Joint Publication 5-0, Joint Planning, Washington, D.C.: Joint Chiefs of Staff, December 1, 2020.

Kolenda, Chris, Christopher Rogers, and Rachael Reid, *The Strategic Costs of Civilian Harm: Applying Lessons from Afghanistan to Current and Future Conflicts*, Washington, D.C.: Open Society Foundations, 2016.

Koplow, David A., "Red-Teaming NLW: A Top Ten List of Criticisms About Non-Lethal Weapons," *Case Western Reserve Journal of International Law*, Vol. 47, No. 1, Spring 2015. As of October 5, 2021:
http://scholarlycommons.law.case.edu/jil/vol47/iss1/17

Kung, Jerry J., *Non-Lethal Weapons in Noncombative Evacuation Operations*, Monterey, Calif.: Naval Postgraduate School, December 1999. As of October 5, 2021:
https://calhoun.nps.edu/handle/10945/13445

Lagasca, Ben, Susan LeVine, and Brian Long, "Combatant or Collateral Damage? New Technology Offers Urban Ops Advantage," *ARMY Magazine*, August 2015, pp. 27–29. As of October 5, 2021:
https://jnlwp.defense.gov/Portals/50/Documents/Resources/Publications/Journal_Articles/New_Technology_Offers_Urban_Ops_Advantage.pdf

Laub, John H., *Study of Deaths Following Electro Muscular Disruption*, Washington, D.C.: National Institute of Justice, May 2011. As of October 5, 2021:
https://www.ojp.gov/pdffiles1/nij/233432.pdf

Law, David B., "Defense Dept. Pursuing Next-Generation Nonlethal Weapons," *National Defense*, February 1, 2009. As of October 6, 2021:
https://www.nationaldefensemagazine.org/articles/2009/1/31/2009february-defense-dept-pursuing-nextgeneration-nonlethal-weapons

Law, David B., "Next-Generation Non-Lethal Technologies: State-of-the-Art Non-Lethal Weapons, Munitions, and Devices Are Produced as Part of the U.S. Department of Defense's Non-Lethal Weapons Program," *SPIE*, September 14, 2016. As of October 5, 2021:
https://spie.org/news/6484-next-generation-non-lethal-technologies?SSO=1

Leimbach, Wendell B., Jr., "DoD Intermediate Force Capabilities: Bringing the Fight to the Gray Zone," Joint Intermediate Force Capabilities Office, U.S. Department of Defense, Non-Lethal Weapons Program, information brief, 2019. As of October 5, 2021:
https://jnlwp.defense.gov/Portals/50/Documents/Resources/Presentations/IFCOverviewBrief_ColL_short.pdf?ver=2019-10-08-122535-957

Leimbach, Wendell B., Jr., "DoD Non-Lethal Weapons Program Brief," Fredericksburg, Va., presentation, NDIA Armaments System Forum, June 4, 2019.

Leimbach, Wendell B., Jr., "Non-Lethal Weapons? Will Marines Ever Use This Capability?" *Marine Corps Gazette*, August 2019. As of October 5, 2021:
https://mca-marines.org/magazines/marine-corps-gazette/

Leimbach, Wendell B., Jr., "DoD Intermediate Force Capabilities: Bringing the Fight to the Gray Zone," Joint Intermediate Force Capabilities Office, U.S. Department of Defense, Non-Lethal Weapons Program, March 2020, Distribution A: Approved for Public Release. As of October 4, 2021:
https://jnlwp.defense.gov/Resources/Presentations/

LeVine, Susan, *The Active Denial System: A Revolutionary, Non-Lethal Weapon for Today's Battlefield*, Washington, D.C.: Center for Technology and National Security Policy, National Defense University, 2009. As of October 4, 2021:
https://apps.dtic.mil/dtic/tr/fulltext/u2/a501865.pdf

LeVine, Susan, "Non-Lethal Directed Energy Weapons and the National Defense Strategy," Joint Intermediate Force Capabilities Office, U.S. Department of Defense, Non-Lethal Weapons Program, presentation, February 27, 2018. As of October 4, 2021:
https://jnlwp.defense.gov/LinkClick.aspx?fileticket=34TG9NpPdK8%3D&portalid=50&timestamp=1524590067508

LeVine, Susan, "Beyond Bean Bags and Rubber Bullets: Intermediate Force Capabilities Across the Competition Continuum," *Joint Forces Quarterly*, No. 100, First Quarter 2021, pp.19–24.

LeVine, Susan, and Col (Ret) John Aho, "Determining True Demand Signal for Non-Lethal Capabilities," *PKSOI*, July 2015, pp. 19–20.

LeVine, Susan, and Joseph Rutigliano, Jr., "U.S. Military Use of Non-Lethal Weapons: Reality vs Perceptions," *Case Western Reserve Journal of International Law*, Vol. 47, No. 239, 2015. As of October 4, 2021:
https://scholarlycommons.law.case.edu/jil/vol47/iss1/18

Lieber, Eli, "Mixing Qualitative and Quantitative Methods: Insights into Design and Analysis Issues," *Journal of Ethnographic and Qualitative Research*, Vol. 3, No. 4, 2009, pp. 218–227.

Lovelace, Douglas C. Jr, and Steven Metz, *Nonlethality and American Land Power: Strategic Context and Operational Concepts*, Carlisle, Pa.: Strategic Studies Institute, 1998.

Lumbee Tribe Enterprise, LLC, and KSA Integration, Prepared for the Joint Intermediate Force Capabilities Office, U.S. Department of Defense, Non-Lethal Weapons Program, *Senior Leadership Engagement Final Report*, Pembroke, N.C., January 31, 2019.

Mapp, Katherine, "Promising New Tool Protects Ships, Sailors," Naval Surface Warfare Center Panama City Division, Public Affairs, November 21, 2019. As of April 12, 2021:
https://www.navsea.navy.mil/Media/News/SavedNewsModule/Article/2023638/promising-new-tool-protects-ships-sailors/

Marine Corps Warfighting Laboratory, *Science Fiction Futures; Marine Corps Security Environment Forecast Futures: 2030–2045*, Quantico Va., November 2016. As of December 2, 2021:
https://www.mcwl.marines.mil/Portals/34/Documents/FuturesAssessment/Marine%20Corps%20Science%20Fiction%20Futures%202016_12_9.pdf?ver=2016-12-09-105855-733

Martin, M. D., *Non-Lethal Weapons: A Policy Planning Paper*, Washington, D.C.: Office of the Under Secretary of Defense, Policy Planning Division, May 29, 1991.

Martinson, Ryan D., and Andrew Erickson, "War on the Rocks: Re-Orienting American Seapower for the China Challenge," *War on the Rocks*, May 10, 2018. As of October 6, 2021:
https://warontherocks.com/2018/05/re-orienting-american-sea-power-for-the-china-challenge/

Mattis, James, *Summary of the 2018 National Defense Strategy: Sharpening the American Military's Competitive Edge*, Washington, D.C.: U.S. Department of Defense, 2018.

Mezzacappa, Elizabeth, Gordon Cooke, Nasir Jaffery, and Charles Sheridan, "Crowd Characteristics and Management with Non-Lethal Weapons: A Soldier Survey," presentation at the Virtual 82nd Military Operations Research Society Symposium, June 23–24 2014. As of October 6, 2021:
https://apps.dtic.mil/sti/citations/ADA613610

Miles, Matthew B., and Michael A. Huberman, *Qualitative Data Analysis: An Expanded Sourcebook*, 2nd ed., Thousand Oaks, Calif.: Sage Publications, 1994.

Morris, Chris, Janet Morris, and Thomas Baines, "Weapons of Mass Protection: Nonlethality, Information Warfare and Air Power in the Age of Chaos," *Airpower Journal*, Spring 1995. As of October 6, 2021:
https://www.hsdl.org/?abstract&did=439933

Morris, Janet, and Chris Morris, *Nonlethality: A Global Strategy*, West Hyannisport, Mass.: Morris & Morris, revised in 2009. As of October 6, 2021:
https://web.archive.org/web/20120305103649/http://www.m2tech.us/images/upload/Nonlethality-A%20Global%20Strategy.pdf

Morris, Lyle J., Michael J. Mazarr, Jeffrey W. Hornung, Stephanie Pezard, Anika Binnendijk, and Marta Kepe, *Gaining Competitive Advantage in the Gray Zone: Response Operations for Coercive Aggression Below the Threshold of Major War*, Santa Monica, Calif.: RAND Corporation, RR-2942-OSD, 2019. As of October 6, 2021:
https://www.rand.org/pubs/research_reports/RR2942.html

National Academies of Sciences, Engineering, and Medicine, *An Assessment of Illness in U.S. Government Employees and Their Families at Overseas Embassies*, Washington, D.C.: The National Academies Press, 2020. As of October 6, 2021:
https://doi.org/10.17226/25889

National Research Council, *An Assessment of Non-Lethal Weapons Science and Technology*, Washington, D.C.: The National Academies Press, 2003. As of October 6, 2021:
https://doi.org/10.17226/10538

NATO—*See* North Atlantic Treaty Organization.

North Atlantic Treaty Organization, Non-Lethal Weapons Effectiveness Assessment Development and Verification Study, Final Report of Task Group SAS-060, Neuilly, France: NATO, RTO-TR-SAS-060, October 2009.

North Atlantic Treaty Organization, Science and Technology Organization, *Analytical Support to the Development and Experimentation of NLW Concepts of Operation and Employment*, Brussels, Belgium, STO-TR-SAS-094, 2017.

Nelson, John, and Wendell Leimbach, "Workshop 6–Virtual Wargaming. A Virtual Wargame Workshop Based on the Intermediate Force Capability Wargame," SAS-151 Hybrid Threats Wargame held virtually at International Concept Development and Experimentation Conference, October 26–29, 2020. As of October 6, 2021:
https://www.act.nato.int/application/files/1116/0471/0504/2020_icde_report.pdf

Nelson, D. A., M. T. Nelson, T. J. Walters, and P. A. Mason, "Skin Heating Effects of Millimeter Wave Irradiation: Thermal Modeling Results," *IEEE Transactions on Microwave Theory and Techniques*, Vol. 48, 2000, pp. 2111–2120.

Ogawa, James, *Evaluating the U.S. Military's Development of Strategic and Operational Doctrine for Non-Lethal Weapons in a Complex Security Environment*, Monterey, Calif.: Naval Postgraduate School, 2003.

O'Hanlon, Michael, "Better Weapons for Stopping Riots: Investments in Acoustic and Jamming Technology Would Help Cops Deal with Mobs Without Shooting," *Wall Street Journal*, January 10, 2021. As of October 6, 2021:
https://www.wsj.com/articles/better-weapons-for-stopping-riots-11610313575

O'Hanlon, Michael E., "Troops Need Not Shoot in Afghanistan," *Politico*, April 23, 2010. As of July 31, 2020:
https://www.politico.com/story/2010/04/
troops-need-not-shoot-in-afghanistan-036223

O'Hanlon, Michael E., *The Senkaku Paradox: Risking Great Power War over Small Stakes*, Washington, D.C.: Brookings Institution Press, April 2019.

Pettyjohn, Stacie L., and Becca Wasser, *Competing in the Gray Zone, Russian Tactics and Western Responses*, Santa Monica, Calif.: RAND Corporation, RR-2791-A, 2019. As of October 6, 2021:
https://www.rand.org/pubs/research_reports/RR2791.html

Police Executive Research Forum, *2011 Electronic Control Weapon Guidelines*, Washington, D.C.: Office of Community Oriented Policing Services, U.S. Department of Justice, March 2011. As of October 6, 2021:
https://www.policeforum.org/assets/docs/Free_Online_Documents/Use_of_Force/electronic%20control%20weapon%20guidelines%202011.pdf

Reaves, Brian, "Local Police Departments, 2013: Equipment and Technology," Bureau of Justice Statistics, NCJ 248767, July 2015. As of October 6, 2021:
https://www.bjs.gov/content/pub/pdf/lpd13et.pdf

Savitz, Scott, Miriam Matthews, and Sarah Weilant, *Assessing Impact to Inform Decisions: A Toolkit on Measures for Policymakers*, Santa Monica, Calif.: RAND Corporation, TL-263-OSD, 2017. As of July 20, 2020:
https://www.rand.org/pubs/tools/TL263.html

Schmitt, Eric, "U.S. Troops Injured in Syria After Collision with Russian Vehicles," *New York Times*, August 26, 2020.

Scott, Richard L., "Nonlethal Weapons and the Common Operating Environment," *ARMY Magazine*, April 2010, pp. 21–26. As of October 6, 2021:
https://www.ausa.org/sites/default/files/FC_Scott.pdf

Scott, Richard L., "Apt Violence," *Military Police*, No. 19-11-1, 2019.

Sedberry, Keith, and Shannon Foley, "Modular Human Surrogate for Non-Lethal Weapons (NLW)," *Defense Systems Information Analysis Center*, Vol. 6, No. 1, Winter 2019.

Shear, Michael D., "Border Officials Weighed Deploying Migrant 'Heat Ray' Ahead of Midterms," *New York Times*, August 26, 2020.

Simiscalchi, Joseph, *Non-Lethal Technologies: Implications For Military Strategy*, Maxwell Air Force Base, Ala.: Center for Strategy and Technology, 1998.

Stoudt, David, "South China Sea Questions: Could Speed-of-Light Weaponry Transform Gray Zone Competition?" *National Interest Newsletter*, April 18, 2020. As of October 6, 2021: https://nationalinterest.org/print/blog/buzz/south-china-sea-questions-could-speed-light-weaponry-transform-gray-zone-competition

Straub, Dan, and Hunter Stires, "Littoral Combat Ships for Maritime COIN," *U.S. Naval Institute Proceedings*, Vol. 147/1/1,415, January 2021. As of October 6, 2021: https://www.usni.org/magazines/proceedings/2021/january/littoral-combat-ships-maritime-coin

Swanson, Anna, Edward Wong, and Julian E. Barnes, "U.S. Diplomats and Spies Battle Trump Administration over Suspected Attacks," *New York Times*, October 19, 2020.

Swett, Charles F., *Strategic Assessment: Non-Lethal Weapons*, Washington, D.C.: Office of the Assistant Secretary of Defense for Special Operations and Low-Intensity Conflict, November 9, 1993.

Swett, Charles F., and Dan Goure, *Non-Lethal Weapons Policy Study Final Report*, Washington, D.C.: Center for Strategic and International Studies, February 5, 1999.

Tafolla, Tray J., David J. Trachtenberg, and John A. Aho, "From Niche to Necessity: Integrating Nonlethal Weapons into Essential Enabling Capabilities," *Joint Forces Quarterly*, No. 66, Third Quarter, 2012.

Taw, Jennifer, David Perselling, and Maren Leed, *Meeting Peace Operations' Requirements While Maintaining MTW Readiness*, Santa Monica, Calif.: RAND Corporation, MR-921-A, 1998. As of October 6, 2021: https://www.rand.org/pubs/monograph_reports/MR921.html

Trachtenberg, David, and William E. Malone, "Non-Lethal Weapons: The Right Tools for the Job," *Jane's Defense Weekly*, February 19, 2009. As of October 6, 2021: https://jnlwp.defense.gov/DesktopModules/ArticleCS/Print.aspx?PortalId=50&ModuleId=9981&Article=577799

Trevithick, Joseph, "Navy to Add Laser Weapons to At Least Seven More Ships in the Next Three Years," *The Drive*, July 2020. As of October 6, 2021: https://www.thedrive.com/the-war-zone/34663

Trump, Donald J., *National Security Strategy*, Washington, D.C., December 2017. As of October 6, 2021:
https://trumpwhitehouse.archives.gov/wp-content/uploads/2017/12/NSS-Final-12-18-2017-0905.pdf

Tucker, Patrick, "The US Military Is Making Lasers Create Voices out of Thin Air," *Defense One*, March 2, 2018. As of April 12, 2021:
https://www.defenseone.com/technology/2018/03/us-military-making-lasers-create-voices-out-thin-air/146824/

U.S. Air Force, *Active Denial System Maritime Operational Assessment Status Report*, AFOTEC/XO, 2006.

U.S. Air Force, *Air Force Requirements for Operational Capabilities Council Memorandum: Active Denial System (ADS) Capability Development Document (CDD)*, HQ USAAF/A5R, January 13, 2009.

U.S. Department of Defense, *Irregular Warfare: Countering Irregular Threats, Joint Operating Concept*, Version 2.0, Washington, D.C., May 17, 2010.

U.S. Department of Defense, *CRM for IW Mission Analysis, Standardized Comment Matrix Primer*, 2016.

U.S. Department of Defense, *Report on Civilian Casualty Policy: Submitted Pursuant to Section 936 of the National Defense Authorization Act for Fiscal Year 2019*, D-B3BEDF4, 2019. As of October 6, 2021:
https://media.defense.gov/2019/Feb/08/2002088175/-1/-1/1/dod-report-on-civilian-casuality-policy.pdf

U.S. Department of Defense, Joint Chiefs of Staff, *Irregular Warfare Mission Analysis*, J7 Directorate for Joint Force Development, SO/LIC, November 2020.

U.S. Department of Defense, Joint Chiefs of Staff, "JS Form 136: Planner Level Review of the Irregular Warfare Mission Analysis," J7 Directorate for Joint Force Development, November 2020.

U.S. Department of Defense, Joint Chiefs of Staff, Deployable Training Division, *Integration of Lethal and Nonlethal Actions, Third Edition*, Washington, D.C.: J7 Directorate for Joint Force Development, 2016.

U.S. Department of Defense, Joint Requirements Oversight Council, *Initial Capabilities Document for Counter Personnel Joint Non-Lethal Effects and Initial Capabilities Document for Counter Materiel Joint Non-Lethal Effects*, Quantico, Va.: JROCM 060-09, April 7, 2009.

U.S. Department of Defense, Office of General Counsel, *Department of Defense Law of War Manual*, Washington, D.C., June 2015, updated December 2016.

U.S. Department of the Navy, Joint Integrated Product Team, *JIPT 20-01 Summary of Conclusions*, March 4, 2020.

U.S. Government Accountability Office, *Defense Management: DOD Needs to Improve Program Management, Policy, and Testing to Enhance Ability to Field Operationally Useful Non-Lethal Weapons*, Washington, D.C., GAO-09-344, April 2009. As of April 30, 2021:
https://www.gao.gov/products/gao-09-344

U.S. Marine Corps, Futures Directorate, *2015 Marine Corps Security Environment Forecast: Futures 2030–2045*, Quantico, Va., 2015. As of October 6, 2021:
https://www.mcwl.marines.mil/Portals/34/Documents/2015%20MCSEF%20-%20Futures%202030-2045.pdf

Walker, Jacob, *Nonlethal Small-Vessel Stopping with High-Power Microwave Technology*, Dahlgren, Va.: Naval Surface Warfare Center, Dahlgren Division, corporate communication, 2012. As of October 5, 2021:
https://apps.dtic.mil/dtic/tr/fulltext/u2/a559057.pdf

Wasser, Becca, Jenny Oberholtzer, Stacie L. Pettyjohn, and William Mackenzie, *Gaming Gray Zone Tactics: Design Considerations for a Structured Strategic Game*, Santa Monica, Calif.: RAND Corporation, RR-2915-A, 2019. As of October 5, 2021:
https://www.rand.org/pubs/research_reports/RR2915.html

Wingenbach, Karl E., and Donald G. Lisenbee, Jr., "'Deconfusing' Lethal and Kinetic Terms," *Joint Force Quarterly*, No. 46, Third Quarter 2007.